O&B
MATHS BANK
4

K. J. Dallison M.A.

J. P. Rigby B.Sc.

Oliver & Boyd

Illustrated by David T. Gray and Tom Reid

Oliver & Boyd
Robert Stevenson House
1–3 Baxter's Place
Leith Walk
Edinburgh EH1 3BB

A Division of Longman Group Ltd.

ISBN 0 05 003157 0

First published 1980
Second impression 1981

Printed in Hong Kong by
Wilture Enterprises (International) Ltd

Contents

Foreword

I first came across the authors' enthusiasm as teachers ten years ago, when they founded our local mathematics association for sixth formers and invited me to become its first president.

All mathematics teachers who use one of the newer types of school syllabus will recognise the usefulness of these books. The introduction of any new syllabus should always be backed up by a wealth of problems, not only for the teacher's benefit, but also to provide side paths along which the pupils can explore and play. This is especially true in the case of the more gifted, and especially in the mixed ability classes of today, in which the more gifted must necessarily be expected to work more on their own.

Learning (that is assimilating and memorising) any syllabus can be a tiresome business, and in mathematics at any level there are always opportunities for indulging in the more challenging and exciting tasks of discovery and creativity as well. These require reflection alone, and some form of guidance. There is no better guidance than well chosen problems that appeal to the intuition and focus the imagination, and through which the student can recreate his or her own mathematics. Such self-discovery leads to a much deeper understanding, and a confidence in the subject, which the student will never forget and upon which he or she can build further.

E. C. ZEEMAN, F.R.S.
University of Warwick

Preface

These books are rather different from the usual mathematics texts in that they contain almost no teaching material. They do, however, contain a wealth of questions, covering the modern and traditional mathematics required by 'O' level and C.S.E. syllabuses in modern mathematics.

Maths Bank can be used to supplement any existing course in modern mathematics but it can also be used as a course book in its own right, leaving the teacher free to instruct in his own way.

The questions are designed to cater for a wide range of ability: each section begins with easier questions; harder and deliberately wordy questions are starred.

Included in this book are several 'O' level examination questions, and for permission to reproduce these the authors would like to thank the Oxford Delegacy of Local Examinations and the Oxford and Cambridge Schools Examination Board (MEI St Dunstan's).

The authors also wish to express their gratitude to Miss P. M. Southern, Mr E. P. Willin and other past and present members of the mathematics staff at Rugby High School for writing questions and supplying answers; to Mr M. E. Wardle, head of the Department of Mathematics at Coventry College of Education for acting as adviser on difficult points; and to Miss D. M. Linsley, former headmistress of Rugby High School, without whose foresight in allowing the school to change to modern mathematics in 1963 these books would never have been written.

K. J. DALLISON
J. P. RIGBY
Rugby High School

1 Groups

1A Revision of Modular Arithmetic and Closure

1 The table shows the operation of
addition for the set of integers (mod 4).
Make similar tables for addition in mod 5
and mod 7. From your tables solve the
following equations in each system:

+	0	1	2	3
0	0	1	2	3
1	1	2	3	0
2	2	3	0	1
3	3	0	1	2

  $a)\ x+3=2$ $b)\ 1-x=2$

2 The table shows the operation of
multiplication for the set of integers (mod 4).
Make similar tables for multiplication in
mod 6 and mod 7.
From your tables solve the following
equations in each system. Where this is not
possible say so. $a)\ 2\times y=2$ $b)\ 2\times y=3$

×	0	1	2	3
0	0	0	0	0
1	0	1	2	3
2	0	2	0	2
3	0	3	2	1

3 How many different numbers are there in mod 6 arithmetic?

4 Write down 5 numbers in ordinary arithmetic which are equivalent to 4 in mod 6
arithmetic.

5 Show that $(2+3)+4=2+(3+4)$ in $a)$ mod 5 arithmetic $b)$ mod 6 arithmetic.

6 Find the values of $(2\times3)\times4$ and $2\times(3\times4)$ in $a)$ mod 5, $b)$ mod 6 arithmetic. What
can you say about addition and multiplication for these two sets of numbers?

7 Solve the following equations in mod 8 arithmetic. Draw up operation tables to
help you find the answers.

  $a)\ 4+x=1$ $b)\ (5+x)+2=3$ $c)\ 3\times y=5$ $d)\ (3\times y)+2=1$ $e)\ x^2=1$

8 Solve the equations of question 7 in mod 9. Where this is not possible, say so.

9 Is the set of numbers in the table in question 1 closed to the operation of addition
(i.e. is the sum of any pair of numbers in the set also a member of the set)? Are the
sets of numbers in the tables you constructed closed to addition?

10 Are the sets of numbers in the table in question 2 and in the tables you
constructed closed to multiplication?

11 If L is rotation through $120°$, M is rotation
through $240°$ and I is the identity transformation,
complete the following table for the shape shown:

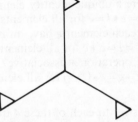

followed	*1st operation*		
by	*I*	*L*	*M*
2nd **I**			
operation **L**		*M*	
M			

Is the set closed to the operation 'followed by' (i.e. is the 'product' of any two members
of the set also a member of the set)?

12 *ABCD* is a rhombus. If *V* is reflection in *AC*, *H* is reflection in *BD*, *O* is rotation through 180° about the centre and *I* is the identity transformation, complete the following table for *ABCD*:

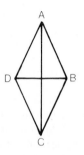

followed		1st operation			
by		*I*	*V*	*H*	*O*
	I				
2nd	*V*				
operation	*H*				
	O				

Do these four operations form a closed set?

1B   Groups

1 *ABCD* is a rectangle. *V* is reflection in the vertical line of symmetry, *H* is reflection in the horizontal line of symmetry, *O* is rotation through 180° about the centre and *I* is the identity transformation.

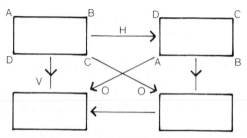

Complete the above diagram to show the different positions of the rectangle under the different transformations.

Complete the table for the first operation followed by the second. ∗ means 'followed by'.

∗		1st operation			
		I	*V*	*H*	*O*
	I				
2nd	*V*				
operation	*H*				
	O				

a) Do these four operations form a closed set?
b) Is there a unique identity element *I*?
(i.e. $I*e=e*I=e$ for all elements *e* of the set)
c) Does each element *e* have an inverse *e'*?
(i.e. $e*e'=I=e'*e$ for all elements of the set)
d) Is the operation ∗ associative?
(i.e. $(e*f)*g=e*(f*g)$ for all elements *e, f, g* of the set)

If your answer to each of these 4 questions is yes, then the set of operations *I, V, H* and *O* form a group.
Is this set a group?

2 Copy and complete the addition table (mod 8) for the set of numbers {0, 2, 4, 6}.

+	0	2	4	6
0				
2				
4				
6				

a) What is the identity element?
b) What is the inverse of i) 6   ii) 4   iii) 2?
c) Is the set closed under addition (mod 8)?
d) Simplify: $2+(4+6)$ and $(2+4)+6$ (mod 8). Which law does this illustrate?
e) Solve the equations (mod 8): i) $2+x=0$ ii) $x+x=4$ iii) $x+6=2$
f) The above structure forms a group. Is this group Abelian (i.e. are all the operations commutative)?

3 Draw up a table using addition (mod 12) for the set of numbers {0, 3, 6, 9} and use it to answer the following:

a) What is the identity element?
b) Check that $3+(6+9)=(3+6)+9$ and $(9+9)+6=9+(9+6)$
c) Is the set closed under addition (mod 12)?
d) Give the inverses of each of the numbers in the set.
e) Does the structure form an Abelian group?
f) Solve the equations i) $6+x=0$ ii) $x+9=6$ iii) $x+x=6$

4 Draw up a table using multiplication (mod 10) for the set {3, 5, 7, 9} and use it to answer the following:

a) Is the set closed under multiplication (mod 10)?
b) Does the commutative law hold?
c) Does this structure form a group? Give reasons for your answer.

5 Draw up a table using multiplication (mod 8) for the set {1, 3, 5, 7}.

a) What is the identity element?
b) Give the inverse of each element in the set.
c) Simplify: i) $3 \times (5 \times 7)$ and $(3 \times 5) \times 7$ ii) $7 \times (3 \times 3)$ and $(7 \times 3) \times 3$
Does the associative law hold?
d) Is the set closed under multiplication (mod 8)? Does it form a group?
e) Solve the following for a using your table:
   i) $3 \times 5 = a \times 1$ ii) $5 \times a = 7$ iii) $(a \times 3) \times 5 = 1$ iv) $a \times a = 1$

6 Make a table for {1, 2, 4, 8} under multiplication (mod 16). Does this structure form a group? Give reasons for your answer.

7 Repeat question 6 for the following structures:

a) {0, 1, 2, 3, 4} under addition (mod 5)
b) {0, 4, 8} under addition (mod 12)
c) {1, 4, 8} under multiplication (mod 12)

8 Why do the following not form groups?

a) {odd integers} under addition   b) {0, 1, 2, 3...} under multiplication
c) {1, 2, 3} under multiplication (mod 4)   d) {−1, 0, 1} under multiplication

9 Draw up a table for the set {A, B, C, D} under matrix multiplication

where $A=\begin{pmatrix} 1 & 0 \\ 0 & 1 \end{pmatrix}$ $B=\begin{pmatrix} 0 & -1 \\ 1 & 0 \end{pmatrix}$ $C=\begin{pmatrix} -1 & 0 \\ 0 & -1 \end{pmatrix}$ $D=\begin{pmatrix} 0 & 1 \\ -1 & 0 \end{pmatrix}$

Does this structure form a group? Give reasons for your answer. Which transformations do the matrices represent?

10 Draw up a table for the set $\{R, S, T, U\}$ under matrix multiplication

where $R = \begin{pmatrix} 1 & 0 \\ 0 & 1 \end{pmatrix}$ $S = \begin{pmatrix} 0 & 1 \\ 1 & 0 \end{pmatrix}$ $T = \begin{pmatrix} 0 & -1 \\ -1 & 0 \end{pmatrix}$ $U = \begin{pmatrix} -1 & 0 \\ 0 & -1 \end{pmatrix}$.

Does this structure form a group? Give reasons for your answer. Which transformations do these matrices represent?

11 Repeat question 10 for the set $\{R, U, V, W\}$ where

$V = \begin{pmatrix} -1 & 0 \\ 0 & 1 \end{pmatrix}$ and $W = \begin{pmatrix} 1 & 0 \\ 0 & -1 \end{pmatrix}$.

12 The operation $*$ is defined on the set $\{p, q, r, s\}$ by the table:

a) State the identity element.
b) State the inverse of r.
c) Find x such that $(p*x)*s = r$.
d) Find y such that $y*(y*y) = y$.

$*$	p	q	r	s
p	s	p	q	r
q	p	q	r	s
r	q	r	s	p
s	r	s	p	q

1C Subgroups and Isomorphisms

1 a) For the equilateral triangle shown, let R denote a clockwise rotation through $120°$ about O, S denote a clockwise rotation through $240°$ about O, and I denote the identity transformation. Draw separate sketches of the triangle under R and S, labelling the vertices carefully.
Complete the following table under the operation $*$ meaning 'followed by':

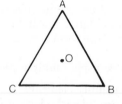

2nd operation $*$	1st operation		
	I	R	S
I			
R			
S			

Does this structure form a group?

b) Draw another equilateral triangle and mark in the three lines of symmetry labelling them l, m and n.

Let L denote reflection in l.
Let M denote reflection in m.
Let N denote reflection in n.

Try to draw up a table for the combination of L, M and N under the operation 'followed by'. What do you find?

c) Draw up a table for the combination of I, R, S, L, M and N under the same operation. Does this structure form a group?

2 You should have found in 1c that the set $\{I, R, S, L, M, N\}$ formed a group under $*$. The three reflections under $*$ did not form a group, but the three rotations in 1a under $*$ did form a group. This group is a *subgroup* of the group formed in 1c.
Look back at the table for 1B question 1. This was the table for the full group of symmetries of the rectangle. Pick out a subgroup of order 2. Identify the two transformations.

3 Complete a similar table for the full group of symmetries of a square where:

I denotes the identity transformation,
A denotes rotation through 90° about its centre *O*,
B denotes rotation through 180° about *O*,
C denotes rotation through 270° about *O*,
H denotes reflection in the horizontal line of symmetry,
V denotes reflection in the vertical line of symmetry,
S denotes reflection in one diagonal,
T denotes reflection in the other diagonal.

Write down and name the group of order 4 which is a subgroup of this group of order 8.

4 Write down the table for the set {0, 1, 2, 3, 4, 5} under addition (mod 6). From this group of order 6 find subgroups of orders 2 and 3.

5 From this table showing the set {*p, q, r, s*} under the operation ∗ answer the following:

a) Does this structure form a group?
b) State the identity element.
c) Find *x* such that $(p*x)*s=r$.
d) Find *y* such that $q*(y*r)=r$.
e) Write down a subgroup of order 2.

∗	*p*	*q*	*r*	*s*
p	*p*	*q*	*r*	*s*
q	*q*	*p*	*s*	*r*
r	*r*	*s*	*p*	*q*
s	*s*	*r*	*q*	*p*

6 In 1B question 5 you were asked to write the table for the set {1, 3, 5, 7} under multiplication (mod 8). Compare that table with the one above. Write down the number which corresponds to each letter. These two groups are *isomorphic*.

7 In 1B question 10, the four transformations formed a group under the given operation. Compare that table with the mod 8 multiplication table which you have just looked at. Write down the corresponding numbers and letters to show that the groups are isomorphic.

8 Draw up the table for the set {2, 4, 6, 8} under multiplication (mod 10). Compare this table with the table for the group of rotation symmetries of the square (part of question 3 above). Write down the corresponding numbers and letters to show that the groups are isomorphic.

9 The tables in questions 6 and 8 showed two types of '4 groups' which were not isomorphic to each other. Every other '4 group', however, is isomorphic to one or other of these two groups. There are, in fact, only two types of '4 groups'.

1st type ∗	*p*	*q*	*r*	*s*
p	*p*	*q*	*r*	*s*
q	*q*	*p*	*s*	*r*
r	*r*	*s*	*p*	*q*
s	*s*	*r*	*q*	*p*

2nd type ∗	*l*	*m*	*n*	*o*
l	*o*	*l*	*m*	*n*
m	*l*	*m*	*n*	*o*
n	*m*	*n*	*o*	*l*
o	*n*	*o*	*l*	*m*

Compare the table of 1B question 11 with the two tables above.
Write a correspondence of letters to show whether it is isomorphic to the first type or the second type.

10 Compare the group tables for {0, 1, 2, 3} under addition (mod 4) and {1, 2, 3, 4} under multiplication (mod 5).

Write down a correspondence between the elements of the two sets, showing that the groups are isomorphic.

11 If $*$ denotes multiplication (mod 10), construct the operation table for the set $\{1, 3, 5, 7, 9\}$ under $*$. Explain why this does not form a group.
Find a subset which does form a group under $*$, stating how each of the necessary conditions is satisfied.

12 Write out the table of the subgroup which you found in question 11 and show that it is isomorphic to one of the groups in question 9.

13 Here are some '3 groups' which have already been studied in this exercise, and also one new one. Are they all isomorphic?

i)

	I	L	M
I	I	L	M
L	L	M	I
M	M	I	L

ii)

	0	4	8
0	0	4	8
4	4	8	0
8	8	0	4

iii)

	I	R	S
I	I	R	S
R	R	S	I
S	S	I	R

iv)

	a	b	c
a	a	b	c
b	b	c	a
c	c	a	b

If your answer was yes, complete the following table showing which elements in the various groups correspond to each other.

	i)	ii)	iii)	iv)
I				
L				
M				

*14 *All* 3 groups are isomorphic, and this can be shown as follows:

Using the letters a, b, c try and make up a '3 group' which is not isomorphic to group iv) of question 13. Call the identity element a, and the table then necessarily starts as shown.

This leaves only 4 blanks which have been numbered 1 to 4. Remembering that there must be only *one* identity and that every element must have only *one* inverse, answer the following questions:

	a	b	c
a	a	b	c
b	b		
c	c		

1	2
3	4

i) b cannot go into spaces 1, 2 or 3. Why not?
ii) c cannot go into spaces 2, 3 or 4. Why not?
iii) The only possible arrangement is therefore:

$$c \quad a$$
$$a \quad b$$

This agrees with table iv) above and confirms that there is only one possible arrangement for a '3 group', i.e. all '3 groups' are isomorphic.

15 List all the '2 groups' you can find in this chapter, including any that are subgroups of larger groups. Are they all isomorphic? Are all '2 groups' isomorphic?

*16 The fact that all '2 groups' are isomorphic can be proved in the same way as in question 14 using a and b and taking a as the identity element. This time after arranging the first row and column there is only one space left, and that cannot be filled by b. Why not?
It must, therefore, be filled by a, i.e. there is only one possible arrangement for a '2 group', so all '2 groups' are isomorphic.

2 Sets

2A Revision

1 $\mathcal{E} = \{$Integers from 6 to 24 inclusive$\}$, $A = \{$Even numbers$\}$, $B = \{$Odd numbers$\}$, $C = \{$Multiples of 3$\}$, $D = \{$Prime numbers$\}$.

a) List the sets *i*) C *ii*) D *iii*) $B \cap C$ *iv*) $B \cap C'$ *v*) $A \cap D$.
b) What do you notice about a)*ii*) and a)*iv*)? Would this be true for every set of integers? Would your answer to a)*v*) be true for every set of integers?
c) Describe in words $A \cap C$.
d) Complete the statement $A \cup B = \underline{\hspace{2em}}$.

2 $\mathcal{E} = \{$Integers from 10 to 30 inclusive$\}$, $P = \{$Square numbers$\}$, $Q = \{$Factors of 144$\}$, $R = \{$Multiples of 5$\}$.

a) List the sets *i*) P *ii*) Q *iii*) $P \cap Q$ *iv*) $P \cup R$.
b) Find the value of $n(R)$.
c) Draw a Venn diagram to show $\mathcal{E}$, P, Q and R.
d) Complete the statement $R \subset \underline{\hspace{2em}}$.

3 $\mathcal{E} = \{$Integers from 1 to 15 inclusive$\}$, $A = \{1, 5, 9, 14\}$, $B = \{2, 3, 4, 5, 6\}$, $C = \{1, 5, 6, 7, 8, 13\}$.

a) List the sets *i*) $A \cap C'$, *ii*) $(B \cap C) \cup A$, *iii*) $A \cap B \cap C$.
b) State whether the following are true or false:

i) $n(B) = 5$ *ii*) $5 \in B$ *iii*) $n(A \cup B) = 9$.

Draw a Venn diagram to show $\mathcal{E}$, A, B and C and all the elements of $\mathcal{E}$.

4 $\mathcal{E} = \{$Integers between 30 and 43$\}$, $P' = \{32, 33, 34, 35, 40, 42\}$, $Q' = \{31, 34, 35, 38, 39, 40\}$, $R' = \{31, 32, 37, 38, 40\}$.

a) List the sets *i*) $P \cap Q \cap R$ *ii*) $(P \cup Q \cup R)'$ *iii*) $P' \cap Q' \cap R'$.
b) Draw a Venn diagram to show $\mathcal{E}$, P, Q and R and all the 12 elements of $\mathcal{E}$.

5 $\mathcal{E} = \{x$ is an integer and $x : 1 \leqslant x \leqslant 10\}$, $A' = \{1, 2, 4, 5, 6, 8, 9, 10\}$, $B' = \{2, 3, 4, 5, 7, 8, 9, 10\}$, $C' = \{3, 4, 7, 8\}$.

a) Draw a Venn diagram to show $\mathcal{E}$, A, B and C.
b) State whether the following are true or false:

i) $6 \notin C$ *ii*) $B \subset C$ *iii*) $B \cap C = C$ *iv*) $A \subset C'$ *v*) $B' \cup A' = \{4, 8\}$.

6 $\mathcal{E} = \{$Letters in the words EQUILATERAL TRIANGLES$\}$, $W = \{$Letters in the words EQUAL ANGLES$\}$, $X = \{$Letters in the word LINES$\}$, $Y = \{$Letters in the word SQUARE$\}$, $Z = \{$Letters in the word ALTERNATE$\}$.

a) List the sets *i*) W' *ii*) $W \cap X$ *III*) $Y \cup Z'$ *iv*) $X' \cap Z$
v) $(W \cup Z)' \cup Y$ *vi*) $(W \cap X) \cup (Y \cap Z)$.

7 If $Z = \{P, R, I, S, M\}$ and $Z' = \{D, A, Y\}$, find $\mathcal{E}$. If $W' = \{S, P, R, A, Y\}$, list the set W.

8 Write four subsets of the set $\{x, y, z\}$. How many are there altogether?

9 The universal set consists of positive integers not greater than 12.

$X = \{x : x^2 \geqslant 50\}$, $Y = \{x : 2x + 7 \leqslant 26\}$, $Z = \{x : 7 < x(x+1) < 70\}$.
List the sets *a)* X *b)* Y' *c)* Z *d)* $X \cap Y$ *e)* $Y \cap Z'$ *f)* $Z \cup Y'$
g) $(X \cup Z)'$ *h)* $X' \cup Z'$.

10 $P = \{x : -1 < x \leqslant 6\}$ and $Q = \{y : -4 \leqslant y < 3\}$. If x and y are both integers, list $P \cap Q$.

The Power of a Set

11 $A = \{(x, y) : x + y = 5$ and x and y are integers$\}$. Calculate $n(A)$ if neither x nor y is negative.

12   Given that $n(R) = 10$, $n(S) = 13$ and $n(R \cup S) = 18$, find $n(R \cap S)$.

13   If $n(\mathscr{E}) = 20$, $n(X) = 14$ and $n(Y) = 8$, and X and Y are subsets of $\mathscr{E}$, find the greatest and least possible values of $n(X \cup Y)$.

14   If $n(\mathscr{E}) = 25$, $n(P) = 11$ and $n(Q) = 18$, and P and Q are subsets of $\mathscr{E}$, find the greatest and least possible values of $n(P \cap Q)$.

15 F and G are subsets of $\mathscr{E}$ and $n(\mathscr{E}) = 32$, $n(F) = 20$ and $n(G) = 16$.
Find the maximum value of *a)* $n(F \cup G)'$ *b)* $n(F \cap G)'$.

16 A, B and C are three sets, and the number of elements in each of the subsets is shown in the Venn diagram. The universal set $\mathscr{E} = A \cup B \cup C$.
Calculate

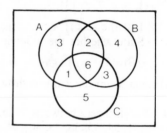

a) $n(A)$ *b)* $n(B \cap C)$ *c)* $n(A \cup C)$
d) $n(A' \cap B')$ *e)* $n(A \cup B)'$ *f)* $n(A \cap B \cap C)'$.

17   If the universal set consists of the first 50 positive integers, and $P = \{x : x$ is divisible by $7\}$, $Q = \{x : x$ is divisible by $14\}$ and $R = \{x : x$ is divisible by $21\}$, find the values of   *a)* $n(P)$ *b)* $n(Q)$ *c)* $n(R)$.

Draw a Venn diagram to show the relationship of P, Q and R. Do not write in the elements.
Describe $P \cap Q \cap R$ in words.

18   The numbers in the Venn diagram show the number of elements in the subsets. If $\mathscr{E} = A \cup B \cup C$ and $n(\mathscr{E}) = 46$, calculate the value of x. Find also

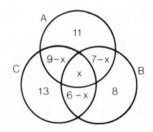

a) $n(A \cup B)'$ *b)* $n(B \cap C')$

2B   More Revision

1   In the universal set $\{31, 32, 33, 34, 35, 36, 37, 38, 39\}$, X is the set of multiples of 3, Y is the set of prime numbers and Z is the set of square numbers. List the members
of   *a)* X *b)* Y *c)* Z *d)* $X' \cap Y$ *e)* $Y' \cap Z'$ *f)* $(X \cap Z)'$
g) $Y \cup (X \cup Z)'$.

2 In the universal set of positive integers not greater than 15, the sets X and Y are given by $X = \{x : x^2 \geqslant 60\}$, $Y = \{x : 5 \leqslant x < 10\}$. List the elements of each of the following sets   a) X b) Y c) $X \cup Y$ d) $X \cap Y$ e) $X' \cap Y$ f) $X' \cup Y'$
g) repeat the answers for c) and f) in the form in which Y is given.

3 $\mathscr{E} = \{x : -10 \leqslant x \leqslant 10\}$, $P = \{x : -3 \leqslant x < 5\}$, $Q = \{x : 0 \leqslant x \leqslant 8\}$, $R = \{x : -7 < x < 1\}$.

List the elements of the following sets   a) $P \cap Q$ b) $P \cap R$ c) $Q \cap R$
d) $P' \cap R$ e) $Q' \cap R$ f) $(P \cup Q)'$ g) $(P \cup Q \cup R)'$

4   Set $A = \{$Divisors of $10\} = \{1, 2, 5, 10\}$

   a) List the members of i) Set $B = \{$Divisors of $12\}$ ii) Set $C = \{$Divisors of $15\}$.
   b) If A, B and C are subsets of the universal set of positive integers not greater than 15, find two members of $(A \cup B \cup C)'$.
   c) Draw a Venn diagram to show the relationship between A, B and C and put the elements in the appropriate regions.

5   The Venn diagram shows the sets A, B and C and the number of elements in each subset.
Write down the number of elements in each of the following:

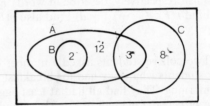

   a) $A \cup B$ b) $(A \cup C) \cap B'$ c) $B \cup C$.

   d) If $n(\mathscr{E}) = 31$, find the value of $n(A \cup B)'$.

6   In a group of 50 children, $B = \{$Children who ride bicycles$\}$ and $P = \{$Children who ride ponies$\}$. If $n(B) = 48$ and $n(P) = 9$, find the maximum and minimum values of $n(B \cap P)$.
If $n(B \cap P) = x$ and $n(B \cup P)' = y$, write down an equation connecting x and y.

7   The universal set $\mathscr{E} = \{x : x$ is an integer and $22 < x \leqslant 37\}$, $V = \{$Even numbers$\}$, $H = \{$Multiples of $3\}$ and $P = \{$Prime numbers$\}$. Draw a Venn diagram and write all the elements in the appropriate spaces.

Describe   a) Set A where $V \cap H = A$
          b) Set B where $H \cap V' = B$
          c) Set C where $V \cup H = C$.

8 $\mathscr{E} = \{x : x$ is an integer and $-5 \leqslant x \leqslant 5\}$, $A' = \{x : 2x > 3\}$, $B' = \{x : x + 7 \leqslant 6\}$, $C' = \{x : 2 \leqslant x + 4 \leqslant 8\}$.

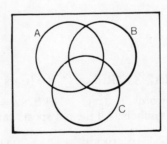

   a) Copy the Venn diagram and write all the elements in the appropriate spaces.
   b) List the members of $A \cap B \cap C'$.
   c) Find the value of $n[(A \cup C)' \cap B]$.
   d) Write in set symbols the three sets which are empty.

9   If $n(\mathscr{E}) = 45$ and P and Q are subsets of $\mathscr{E}$ such that $n(P) = 22$ and $n(Q) = 17$, calculate the minimum value of a) $n(P' \cap Q')$ b) $n(P \cap Q')$.

10   In a group of 100 science students, each studied at least one of the three subjects physics, chemistry and biology.
12 studied physics only; 20 studied physics but not chemistry; 43 studied physics and

biology; 64 studied physics; 68 studied biology; 41 studied chemistry and biology.
If P, C and B represent the sets of students who studied physics, chemistry and biology respectively, show P, C and B on a Venn diagram.

   a) Describe in words *i*) $P \cap C \cap B$ *ii*) $P' \cap B'$.
   b) Calculate the number of students in each of the sets in *a*).
   c) Calculate also the number of students who studied *i*) one and only one of these subjects, *ii*) two and only two of these subjects.

11 In a College of Further Education, each one of an intake of 50 language students studied at least one of these languages—French, German and Spanish. 26 studied French and German, and of these 5 studied Spanish as well. 2 studied German and Spanish but not French. 4 studied German only. There were 40 students in the French class and 12 in the Spanish class.

   a) How many studied both French and Spanish?
   b) Denoting by F, G and S the sets of students who studied French, German and Spanish respectively, describe in words $F \cap G' \cap S$.
   c) Find $n(F \cap G' \cap S)$.   *d*) Find also $n(F' \cap G' \cap S)$.

*12 Two college students asked a number of other students what they had for breakfast that morning. They had all had at least egg, bacon or cereal. One half of those who had an egg had bacon as well. A third of those who had bacon and eggs had cereal also. One quarter of those who had an egg also had cereal but no bacon. The number who had bacon only was twice the number who had cereal and bacon but no egg. The number who had cereal only was the same as the number who had all three.
Let C, B and E be the sets of people who had eaten cereal, bacon and egg respectively. Let $n(E) = 12x$ and $n(C \cap B \cap E') = y$. Show in a Venn diagram, in terms of x and y, the numbers of people in each region. If altogether 96 people were questioned, write down an equation in x and y. If 8 of them had bacon only, find how many had cereal only and how many only had egg.

2C

1 If $\mathscr{E} = \{\text{Cars}\}$, $A = \{\text{Red cars}\}$, $B = \{\text{Cars registered this year}\}$ and $C = \{\text{Sports cars}\}$,

   a) write in words: *i*) $A \supset C$ *ii*) $B \cap A = \phi$ *iii*) $C \subset B'$
   b) write in symbols: All the red sports cars were registered this year.

2 If $\mathscr{E} = \{\text{Mice}\}$, $W = \{\text{White mice}\}$, $L = \{\text{Mice with long tails}\}$,
$P = \{\text{Mice with pink ears}\}$, write in set language

   a) All white mice have long tails.
   b) Some white mice have pink ears.

Write sentences to express the following statements given in symbols:

   c) $L \cap P = P$ *d*) $L \cap W' \neq \phi$.

18

3 If $\mathcal{E}=\{$Houses$\}$, $D=\{$Detached houses$\}$, $G=\{$Houses with garages$\}$, $T=\{$Two-storey houses$\}$, $Y=\{$Houses built this year$\}$, write sentences which express the following statements:

a) $G\supset Y$ b) $(D\cup Y)\subset G$ c) $T\cap G'=\phi$.

Write in symbols: d) Some of the houses built before this year have garages.

e) No detached houses with more or less than two storeys have been built this year.

4 In this question $\mathcal{E}=\{$Men$\}$, $B=\{$Bald-headed men$\}$, $F=\{$Men over 50 years of age$\}$, $S=\{$Single men$\}$, $T=\{$Tall men$\}$. Write sentences which express the following statements:

a) $S\subset B'$ b) $F\cap T\neq\phi$ c) $T\cup S=S$ d) $(T\cap B)\subset F$

5 In this question $\mathcal{E}=\{$Books$\}$, $T=\{$Text books$)$, $P=\{$Paper-backed books$\}$, $L=\{$Library books$\}$, $I=\{$Books with pictures$\}$.

a) Express the following sentences in symbols:
i) No paper-backed books have pictures.
ii) Some library books are text books.
iii) No text books in a library have pictures.
b) Write a sentence to express the following statement: $P\cap L\neq\phi$.

6 If $\mathcal{E}=\{$Pupils in the school$\}$, $G=\{$Girls$\}$, $U=\{$Pupils under 14$\}$, $F=\{$Members of the football team$\}$, write in symbolic form:

a) Some girls are under 14.
b) All the members of the football team are boys.
c) No boy or girl of 14 or over plays in the football team.

Draw a single Venn diagram to illustrate these three statements.

7 Let the universal set be the set of flowers, P be the set of spring flowers, B be the set of blue flowers and S be the set of flowers with a scent. Write each of the following in set language:

a) All spring flowers are scented.
b) Some blue flowers have no scent.
c) No spring flowers are blue.

Draw a single Venn diagram to illustrate these three statements.

8 If $\mathcal{E}=\{$Boys$\}$, $F=\{$Fat boys$\}$, $G=\{$Greedy boys$\}$, $T=\{$Tall boys$\}$, write the following statements in set symbols:

a) Some tall boys are not fat.
b) Some greedy boys are tall and fat.
c) All fat boys are either tall or greedy or both.

9 $\mathcal{E}=\{$Animals$\}$, $D=\{$Dogs$\}$, $B=\{$Black animals$\}$, $L=\{$Animals with long tails$\}$.

a) Write the following statement using set symbols: Some dogs have long tails.
b) Write a sentence to express the following statement: $B\cap L=\phi$.
c) Copy the Venn diagram and label the sets B, D and L. Shade the region representing black dogs.

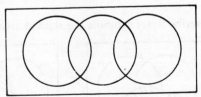

Note Some of the sets in 1–9 were not well defined and were therefore (strictly) not sets.

10 If $\mathscr{E}=\{$Positive integers$\}$, $A=\{$Multiples of 2$\}$, $B=\{$Multiples of 3$\}$, $C=\{$Multiples of 30$\}$, express in set symbols the statement 'All positive integers which are multiples of 30 are also multiples of 2 and 3.'

11 All the members of one class had taken examinations in English, German, French, mathematics and biology. E, G, F, M and B represent the sets of students who had passed in these five subjects respectively. Express each of the following statements in set symbols:

a) All members of the class had passed in English or mathematics or both.
b) No one who had passed in German had failed in French.
c) Everyone who had passed in biology had also passed in mathematics.
d) Everyone who failed in biology had also failed in German.

12 If $\mathscr{E}=\{$Triangles$\}$, $A=\{$Triangles with at least one angle of 60°$\}$, $B=\{$Triangles with at least one line of symmetry$\}$

a) name the triangles defined by *i)* $A\cap B$ *ii)* $A'\cap B$.
b) state which of these sets could contain a right-angled triangle

i) $A\cap B'$ *ii)* $A\cap B$ *iii)* $A'\cap B$ *iv)* $(A\cup B)'$

13 If $\mathscr{E}=\{$Quadrilaterals$\}$, $R=\{$Quadrilaterals with at least one line of symmetry$\}$, $S=\{$Quadrilaterals with at least one pair of parallel sides$\}$, draw a member of:

a) $R'\cap S'$ *b)* $R'\cap S$ *c)* $R\cap S'$.

14 Draw a Venn diagram to show the relationship between the following sets:
$\mathscr{E}=\{$Quadrilaterals$\}$, $C=\{$Cyclic quadrilaterals$\}$, $P=\{$Parallelograms$\}$, $S=\{$Squares$\}$.

15 If $\mathscr{E}=\{$Quadrilaterals$\}$, $L=\{$Quadrilaterals with at least one pair of parallel sides$\}$, $M=\{$Quadrilaterals with at least one pair of opposite sides equal$\}$, $N=\{$Quadrilaterals with opposite angles supplementary$\}$, name a member of each of:

a) $L\cap M$ *b)* $M\cap N$ *c)* $L'\cap M'\cap N$.

16 If $\mathscr{E}=\{$Triangles$\}$, $R=\{$Right-angled triangles$\}$, $S=\{$Triangles with one obtuse angle$\}$, $T=\{$Triangles with two equal angles$\}$, draw a Venn diagram to show the relationship between the sets.
Draw, if possible, a member of each of the following sets:

a) $R\cap T$ *b)* $R\cap S$ *c)* $S\cap T$ *d)* $R'\cap S'\cap T$.

2D

1 In each of the following cases, draw Venn diagrams to show the sets P and Q, which are subsets of the same universal set:

a) $P\cup Q=P$ *b)* $P\cap Q=P$ *c)* $P\cap Q=\phi$.

2 Describe in set symbols the shaded areas shown in the following diagrams:

a) 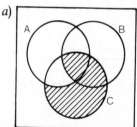   *b)* 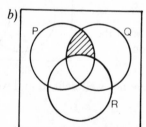   *c)*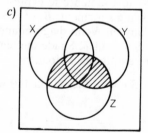

3 Three sets P, Q and R are subsets of a universal set and $P \cap Q \neq \phi$, $R \subset Q$, $R \cap P = \phi$.

By drawing Venn diagrams and shading them (or otherwise) say whether the following statements are true or false:

a) $R \subset P'$ b) $P \cap R' = P$ c) $Q \cup R = R$ d) $Q \cap R = R$.

4 Which of the following statements are true for the sets X and Y shown in the diagram?

a) $X \cap Y = Y$ b) $X \cup Y = X$
c) $Y \cap X' = \phi$ d) $X' \cup Y = \mathscr{E}$
e) $Y' \subset X'$ f) $X \cup Y' = \mathscr{E}$

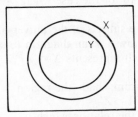

5 Draw separate Venn diagrams for each of a), b), c) and d) and shade the regions stated:

a) $(L \cap M) \cup L'$
b) $(L \cup M) \cap L'$
c) $(L \cap M)' \cap L$
d) $(L \cup M)' \cup L$

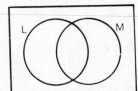

6 Two sets A and B are subsets of the universal set $\mathscr{E}$ such that $A \cap B \neq \phi$, and $A \cup B \neq \mathscr{E}$. Draw separate Venn diagrams to show the possible relationships of the sets. Shade on each, if possible, the region which represents $(A \cup B) \cap B'$.

7 Two sets P and Q are non-zero subsets of the universal set $\mathscr{E}$ such that $P \cap Q \neq \phi$ and $P \cup Q \neq \mathscr{E}$.

By drawing Venn diagrams and shading them appropriately (or otherwise), say whether the following statements must be true, must be false or may be false:

a) $P' \cap Q' = (P \cup Q)'$ b) $P' \cup Q' = (P \cap Q)'$
c) $P' \cup Q' = \mathscr{E}$ d) $P' \cup Q = \mathscr{E}$

If your answer is 'may be false', explain why.

8 Three sets R, S and T are subsets of the universal set $\mathscr{E}$ but not of one another, such that $R \cap S \neq \phi$, $R \cap T = \phi$ and $S \cap T \neq \phi$.

Draw a Venn diagram to show the relationships of the sets.

Shade the region which represents $(R \cup S) \cap T'$.

9 Write a statement in set symbols which defines the relationship between the sets P, Q and R shown in the Venn diagram. Copy the diagram and shade the region $(P \cap Q) \cap R'$.

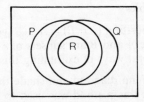

10 Simplify the following expressions for the sets shown in the Venn diagram:

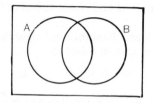

a) $A \cap (A' \cup B)$ b) $A \cup (A' \cap B)$

c) $A' \cap B'$ d) $A' \cup B'$

e) $(A' \cup B') \cup (A \cap B)$

11 Repeat question 10 for the case where $A \cap B = \phi$.

12 If X, Y and Z are subsets of the universal set $\mathscr{E}$ such that $X \cup Y \cup Z \neq \mathscr{E}$, $X \subset Y$ and $X \cap Y \cap Z \neq \phi$, draw a Venn diagram to show the relationship between the sets. Shade the region which represents $X' \cap Y \cap Z'$.

13 In the Venn diagram, shade the region which represents $X \cap Y \cap Z'$.
Write in set notation the relation which must hold between the three sets if the above set is empty.

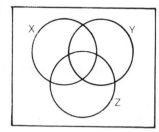

14 A and B are two points in a plane 5 cm apart. Sets of points in the plane are defined as follows: $X = \{P : AP = BP\}$, $Y = \{P : AP = 3 \text{ cm}\}$. Find $n(X \cap Y)$.

15 A and B are two fixed points and sets of points in space are defined as follows: $X = \{P : AP = BP\}$, $Y = \{P : \text{angle } APB = 90°\}$. Describe X, Y and $X \cap Y$.

16 A and B are two fixed points 6 cm apart. A set of points in a plane through A and B is defined as $\{P : \text{angle } APB \geqslant 90°\}$. Describe this set.

17 If $L = \{\text{Points on a line}\}$ and $C = \{\text{Points on the circumference of a circle}\}$

a) describe the line in geometrical terms when i) $n(L \cap C) = 2$ ii) $n(L \cap C) = 1$
iii) $n(L \cap C) = 0$.

✱ b) What can you say about the circle if i) $L = C$ ii) $n(C) = 1$?

18 Using rectangular axes OX and OY, sets of points in the positive quadrant of the plane of the axes are given by $A = \{P : \text{angle } POX \leqslant \text{angle } POY\}$, $B = \{P : x \text{ co-ordinate of } P \leqslant 3\}$, $C = \{P : y \text{ co-ordinate of } P \leqslant 3\}$. Draw a diagram to show $A \cap B \cap C$.

19 $EFGH$ is a square of side 4 cm. A set of points in the plane of the square is given by $Z = \{P : \text{angle } EPF = 45°\}$. Draw a diagram to show Z and describe this set exactly.

2E   Truth Tables

1 Is the proposition $(A \cap B)' = A' \cup B'$ generally true (i.e. true for all possible sets $\mathscr{E}$, A and B where A and B are subsets of $\mathscr{E}$)?
There are several ways of examining this question.

a) Draw a Venn diagram and shade the relevant areas. Name one or two questions in 2D where this method was used to examine similar problems.

b) Define actual sets $\mathscr{E}$, *A* and *B* and list the members of A', B', $(A \cap B)$, etc. See questions 2, 3, 4 below. This method is of limited value. Why?

c) Construct truth tables. See question 5 onwards.

d) Use formal set theory. See 19B.

2 $\mathscr{E} = \{1, 2, 3, 4, 5, 6, 7, 8\}$, $A = \{2, 3, 4, 5\}$ and $B = \{4, 5, 6, 7\}$.

a) Copy and complete the following table which investigates the proposition $(A \cap B)' = A' \cup B'$.

A	*B*	$A \cap B$	A'	B'	$(A \cap B)'$	$A' \cup B'$
2, 3, 4, 5	4, 5, 6, 7	4, 5				

b) The last two columns should contain the same numbers. Does this prove the proposition?

3 Repeat question 2 for the proposition $(A \cup B)' = A' \cap B'$.

4 If the final columns are not the same, this disproves the proposition. So this method offers a quick way of disproving a proposition which you suspect is not true. Taking $\mathscr{E} = \{3, 5, 7, 9, 11, 13\}$, $A = \{5, 9\}$ and $B = \{3, 5, 11\}$, prove that the following propositions are untrue:

a) $A' \cap B = A \cup B'$ *b*) $(A \cup B) \cap A' = B'$.

5 Propositions of the type examined above can also be examined using truth tables. Here is an example for the proposition $(A \cap B)' = A' \cup B'$.

A	*B*	$A \cap B$	A'	B'	$(A \cap B)'$	$A' \cup B'$
1	1	1	0	0	0	0
1	0	0	0	1	1	1
0	1	0	1	0	1	1
0	0	0	1	1	1	1

If an element of $\mathscr{E}$ is in *A*, 1 is entered in column *A*. If it is not in *A*, 0 is entered in column *A*. Similarly with column *B*. So the first line represents an element of $\mathscr{E}$ that is in both *A* and *B*.

a) What does the second line represent?

b) What do the third and fourth lines represent?

c) Would it be possible to add a fifth line, i.e. to find an element of $\mathscr{E}$ that is not contained in lines 1 to 4?

d) How is the third column obtained from the first two columns?

e) How is the fourth column obtained?

f) How are the sixth and seventh columns obtained?

g) The last two columns are identical. Does this prove the proposition?

✳ *h*) Explain simply why the truth table proves or disproves the proposition.

6 Construct truth tables for the following propositions, all of which are true:

a) $A \cup B' = (A' \cap B)'$ *b*) $A \cap B' = (A' \cup B)'$ *c*) $(A' \cup B) \cap A = A \cap B$

7 Construct truth tables for the following propositions, none of which are true:

a) $A \cap B' = A' \cup B$ *b*) $(A \cap B) \cup (A \cup B) = A$ *c*) $A' \cup (A' \cap B) = B'$

8 Using truth tables, investigate the truth or otherwise of the following propositions:

a) $A \cup (A' \cap B) = (A \cup A') \cap (A \cup B)$ b) $A \cap (A \cup B') = A$
c) $(A \cap B) \cup (A \cup B) = A$ d) $(A \cup B) \cap (A \cap B) = B$
e) $(A \cap B') \cup A' = A \cup B'$ f) $(A \cap B') \cup A' = (A \cap B)'$

9 If the proposition concerns three sets A, B and C, then there will be 8 lines in the truth table.
Copy and complete the following table, and find out the truth (or otherwise) of the proposition $A \cap (B \cup C) = (A \cap B) \cup (A \cap C)$.

	A	B	C	$B \cup C$	$A \cap (B \cup C)$	$A \cap B$	$A \cap C$	$(A \cap B) \cup (A \cap C)$
1	1	1	1	1	1	1	1	1
2	1	1	0	1	1	1	0	1
3	1	0	1					
4	1	0	0					
5	0	1	1					
6	0	1	0					
7	0	0	1					
8	0	0	0					

10 Draw up truth tables for the following propositions, all of which are true:

a) $A \cup (B \cap C) = (A \cup B) \cap (A \cup C)$ b) $(A \cap B) \cup (A \cap C) = A \cap (B \cup C)$
c) $A \cap (A \cup B) = A \cup (A \cap C)$

11 Using truth tables, examine the truth (or otherwise) of the following propositions:

a) $(A \cap B) \cup C = (A \cap C) \cup B$ b) $(A \cup B) \cap A = (A \cap C) \cup A$
c) $(A \cup B \cup C)' = A' \cap B' \cap C'$ d) $(A \cup B) \cap (A \cup C) = (A \cap B) \cup (A \cap C)$
e) $(A \cap B)' \cup C = (A' \cup B') \cup C$ f) $(A' \cap B) \cup C = A' \cap (B \cup C)$

12 In the Venn diagram shown, regions 1, 4 and 8 correspond to lines 1, 4 and 8 of the table in question 9. Number the remaining regions.

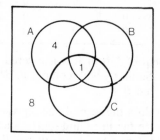

3 Topology

3A Networks and Route Matrices

1 a) Write down the route matrix for the network shown in the diagram.
b) If the routes *AB*, *BC* and *CA* are one way only, in the direction indicated by the letters, rewrite the route matrix.

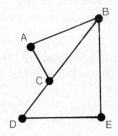

Note Row A should show routes leaving *A*. Column A should show routes arriving at *A*. Similarly with row and column B, C, etc, and with all the matrices in 3A.

2 Write down the direct route matrix for the road network shown in the diagram.

a) Why are the numbers in the leading diagonal all 0's?
b) Are the rows and columns in the matrix interchangeable?

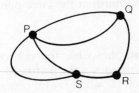

3 Write down the direct route matrix for the road network shown in the diagram. Single arrows indicate one-way roads.
a) Why are the rows and columns in the matrix not interchangeable?
b) Why is there a 1 in the diagonal?

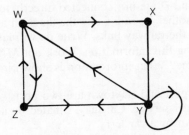

4 Write down the route matrix for the network shown in the diagram. Since there are no arrows to indicate one-way routes, take all routes as two-way.

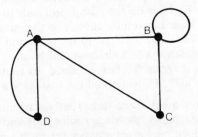

5

$$\begin{array}{c} \\ W \\ X \\ Y \\ Z \end{array} \begin{array}{cccc} W & X & Y & Z \\ \begin{pmatrix} 0 & 1 & 0 & 0 \\ 1 & 0 & 1 & 1 \\ 0 & 1 & 0 & 1 \\ 0 & 1 & 1 & 0 \end{pmatrix} \end{array}$$

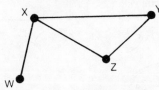

The matrix shows the one-stage routes which link the four points *W, X, Y* and *Z*. Multiply the matrix by itself. The new matrix shows the number of two-stage routes which link the four points (i.e. routes which follow two lines). List all the two-stage routes.

25

6 Write down the route matrix M for the network shown. Work out M^2. How many two-stage routes start and finish a) at the same point, b) at different points?

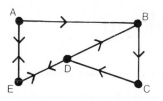

7 Write down the route matrix M for the given network. Work out M^2. List all the two-stage routes.

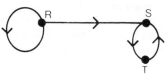

8 Write down the route matrix M for the given network. Work out M^2 and M^3. M^3 shows all the three-stage routes which link P, Q, R and S. List all the three-stage routes which start and finish at the same point.

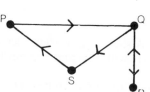

9 The diagram shows bus routes between four towns. Write down the matrix M giving this information. Town A is also connected to each of the other three towns by rail, but B, C and D are not connected directly to one another by rail. Copy the diagram and put in the railway links. Write down matrix

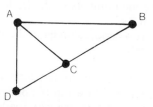

N giving this information. Work out MN and NM. Why are these two matrices not the same? What information is shown in the matrix $M + N$?

10 Alport and Baytown are two ports on the mainland which are linked by a passenger service. Carquay, situated on a large well-inhabited island, is linked to Baytown by a car ferry.
Carquay is also served by a second ferry which leaves Alport, calls at Dridock, a smaller harbour on the island, and sails to Baytown having called at Carquay on the way. It does not return directly to Carquay, Dridock or Alport.
A smaller island is joined to the mainland by a ferry service which starts at Alport and takes passengers to Eelmouth and back.
Draw a diagram to show all these sea routes. Write down the route matrix M. Work out M^2 and M^3, to find all the two-stage and all the three-stage routes.

   a) How many three-stage routes start and finish at the same port?
   b) How many two-stage routes start and finish at the same port?
   c) Which port gives a passenger most variety of choice for a sea trip of not more than three stages?

3B Incidence Matrices

1 Complete the incidence matrix for the network shown in the diagram.

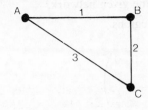

$$\begin{array}{c} \\ \text{Nodes} \end{array} \begin{array}{c} \\ A \\ B \\ C \end{array} \overset{\displaystyle \text{Arcs}}{\overset{\displaystyle 1 \ \ 2 \ \ 3}{\left(\begin{array}{ccc} 1 & & \\ 1 & & \\ 0 & & \end{array}\right)}}$$

2 Write the matrix which shows which nodes are incident on which arcs for each of the given networks.

a)

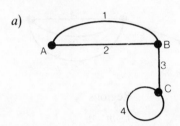

b)

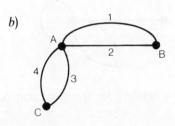

3 Draw a network for the given incidence matrix.
What do you notice about the sum of the numbers a) in each row, b) in each column?

$$\begin{array}{c} \\ \\ \text{Arcs} \end{array} \begin{array}{c} \\ 1 \\ 2 \\ 3 \\ 4 \\ 5 \end{array} \overset{\displaystyle \text{Nodes}}{\overset{\displaystyle A \ \ B \ \ C \ \ D}{\left(\begin{array}{cccc} 1 & 1 & 0 & 0 \\ 0 & 1 & 1 & 0 \\ 0 & 0 & 1 & 1 \\ 0 & 0 & 1 & 1 \\ 0 & 1 & 1 & 0 \end{array}\right)}}$$

4 For the network shown in the diagram, complete the incidence matrix which relates nodes and regions.

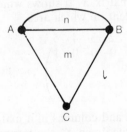

$$\begin{array}{c} \\ \text{Nodes} \end{array} \begin{array}{c} \\ A \\ B \\ C \end{array} \overset{\displaystyle \text{Regions}}{\overset{\displaystyle l \ \ m \ \ n}{\left(\begin{array}{ccc} 1 & 1 & 1 \\ & & \\ & & \end{array}\right)}}$$

5 Write down incidence matrices relating nodes to regions for each of the given networks.

a)

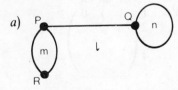

b)

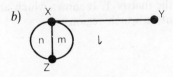

6 Draw a network for the given matrix.

$$\begin{array}{c} \\ \text{Nodes} \end{array} \begin{array}{c} \\ A \\ B \\ C \end{array} \overset{\displaystyle \text{Regions}}{\overset{\displaystyle l \ \ m \ \ n}{\left(\begin{array}{ccc} 1 & 1 & 0 \\ 1 & 1 & 1 \\ 1 & 0 & 1 \end{array}\right)}}$$

7 Complete the matrix which relates arcs
to regions for the given network.

Regions

$$\text{Arcs} \quad \begin{array}{c} 1 \\ 2 \\ 3 \\ 4 \end{array} \begin{pmatrix} l & m & n \\ 1 & 0 & 1 \\ & & \\ & & \\ & & \end{pmatrix}$$

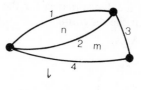

8 Write down matrices relating arcs to regions for each of the given networks.

a)

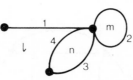

b)

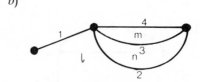

9 Draw a network for the given matrix.

Regions

$$\text{Arcs} \quad \begin{array}{c} 1 \\ 2 \\ 3 \\ 4 \\ 5 \end{array} \begin{pmatrix} l & m & n \\ 1 & 0 & 1 \\ 1 & 0 & 1 \\ 1 & 1 & 0 \\ 0 & 1 & 1 \\ 0 & 1 & 1 \end{pmatrix}$$

10 The diagram shows a network of four nodes connected by five arcs which make
three regions.
Complete the matrix X. It shows which
nodes are incident on which arcs.

$$X = \begin{array}{c} A \\ B \\ C \\ D \end{array} \begin{pmatrix} 1 & 2 & 3 & 4 & 5 \\ 1 & 0 & 0 & 0 & 1 \\ & & & & \\ & & & & \\ & & & & \end{pmatrix}$$

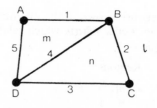

When the rows and columns in a matrix are interchanged, the matrix is said to be
transposed. Write down X', the transpose of X, and work out the product XX'.
Explain the significance of the numbers in the leading diagonal, and of the other terms.

11 For the network given in question 10,
complete the matrix Y. It shows which arcs
are incident on which regions.

$$Y = \begin{array}{c} 1 \\ 2 \\ 3 \\ 4 \\ 5 \end{array} \begin{pmatrix} l & m & n \\ 1 & 1 & 0 \\ & & \\ & & \\ & & \\ & & \end{pmatrix}$$

Complete matrix Z. It shows which nodes
are incident on which regions.
Work out the product XY. What is the
relation between Z and XY?

$$Z = \begin{array}{c} A \\ B \\ C \\ D \end{array} \begin{pmatrix} l & m & n \\ 1 & 1 & 0 \\ & & \\ & & \\ & & \end{pmatrix}$$

28

12 Repeat questions 10 and 11 for the network of three nodes connected by five arcs making four regions.

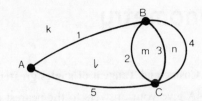

13 Write down Z', the transpose of matrix Z for the network in question 12. Work out the product ZZ'. Explain the significance of the numbers in the leading diagonal, and of the other numbers in the matrix ZZ'. Work out the product $Z'Z$ and explain the significance of the numbers in the leading diagonal and of the other numbers.

14 Repeat question 13 for the network shown, which consists of four nodes connected by five arcs making three regions.

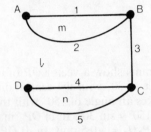

15 Compile X, the node-arc incidence matrix for the network of question 14. Explain how the product XX' differs from the direct route matrix between nodes.

16 The diagram shows the plan of a house where there are three rooms A, B and C and the garden G all round the outside. There are four doors w, x, y and z. Write down a matrix N with four rows, one for each of the spaces and with four columns, one for each of the doors. A '1' in the matrix signifies that the door opens into, or out of, the particular 'space'.

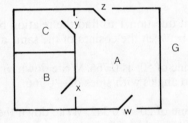

Compile N', the transpose of N and work out the product NN'. Explain the significance of the elements in the leading diagonal and of the other elements in the matrix NN'.

4 Trigonometry

4A Sine, Cosine and Tangent of angles greater than 90°

Note In 4A give your answers to the nearest degree between 0° and 360°.

1

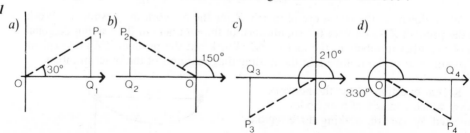

The four diagrams show a vector OP of unit length in different positions. In *a*) OP_1 makes an angle of 30° with the central direction and $P_1Q_1 = 0.5$ unit, so sin 30° = 0·5. In *b*) OP_2 makes an angle of 150° with the central direction and $P_2Q_2 = 0.5$ unit. Therefore sin 150° = sin 30°. In *c*) OP_3 makes an angle of 210° with the central direction and $P_3Q_3 = -0.5$ unit. In *d*) OP_4 makes 330° with the central direction and $P_4Q_4 = -0.5$ unit. Therefore sin 210° = sin 330° = −sin 30°.
Repeat the above to find the relations between the cosines of 30°, 150°, 210° and 330°. (Remember that OQ_2 is in the negative direction and that the length of the vector OP is always positive.)

2 Repeat question 1 to find the relation between the sines of 60°, 120°, 240° and 300°, and also between the cosines of the same angles.

3 The sine of 50° is 0·766. Write down *a*) the other angle which has the same sine, *b*) the two angles with sines of −0·766.

4 The sine of 20° is 0·342. Write down the angles whose sines are either 0·342 or −0·342. State which is which.

5 The cosine of 72° is 0·309. Write down *a*) the other angle which has the same cosine, *b*) the two angles with cosines of −0·309.

6 The cosine of 27° is 0·891. Write down the angles with cosines of 0·891 or −0·891. State which is which.

7 Using tables, write down the values of the following:

a) sin 18°   *b*) sin 243°   *c*) cos 143°   *d*) cos 297°   *e*) cos 218°
f) sin 308°   *g*) sin 126°   *h*) cos 54°   *i*) cos 351°   *j*) sin 260°

8   *a*)

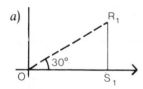

b)

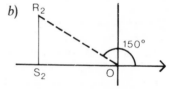

In diagram *a*) OR_1 is making an angle of 30° with the central direction and the tangent of $30° = \dfrac{R_1S_1}{OS_1} = 0·577$ (correct to 3 d.p.).

In diagram b) OR_2 is making an angle of 150° with the central direction and the

tangent of $150° = \dfrac{R_2S_2}{OS_2} = \dfrac{R_1S_1}{-OS_1} = -0·577$.

Draw sketches to show OR making angles of 210° and 330° with the central direction. Find the relation between the tangents of 30°, 150°, 210° and 330°.

9 Repeat question 8 and find the relation between the tangents of 45°, 135°, 225° and 315°.

10 The tangent of 66° is 2·246. Find a) the other angle with a tangent of 2·246, b) the two angles with tangents of $-2·246$.

11 The tangent of 20° is 0·364. Write down the angles with tangents of 0·364 or $-0·364$. State which is which.

12 Using tables, write down the values of the following:

a) tan 22°	b) tan 219°	c) tan 303°	d) tan 232°	e) tan 145°
f) tan 167°	g) tan 80°	h) tan 325°	i) tan 260°	j) tan 351°

4B   Sine, Cosine and Tangent Curves

1 If $y = \sin x$, taking values of x as 0°, 30°, 60°, etc., up to 360°, find the corresponding values of y. Draw a graph of $y = \sin x$. Write down the range of values of x for which y is positive.
For any given value of y, how many values of x are there between 0° and 360°?
Find values of x from your graph for which

a) $y = 0·4$ b) $y = -0·9$ c) $y = 0·65$.

2 Repeat the whole of question 1 for the relation $y = \cos x$.

3 The diagram shows the graph of $y = \tan x$. Using the sketch answer the following questions:

a) For what values of x is y zero?
b) For what values of x is y infinity?
c) For what range of values of x is y positive?
d) For any given value of y, how many values of x are there?
* e) The dotted lines in the figure are called 'asymptotes'. What do they indicate?
* f) There are apparently two totally different values for tan 90°, $+\infty$ and $-\infty$. Can you explain the apparent discrepancy?

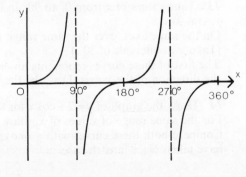

4 From the questions which you have already answered, write down

a) the quadrants in which the sine of an angle is positive,
b) the quadrants in which the cosine is positive,
c) the quadrants in which the tangent of an angle is positive.

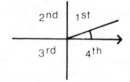

(The diagram shows the four quadrants.)

5 Using the answers to question 4 (and taking all angles between 0° and 360°), find the possible values of A, B.....I in the following. If you are using 3-figure tables, correct the given decimals before you begin.

$\sin A = 0.5592$, $\sin B = -0.9660$, $\cos C = -0.8829$, $\tan D = 1.3270$, $\tan E = 0.6494$, $\cos F = 0.7986$, $\tan G = -2.2460$, $\tan H = -0.3443$, $\cos I = -0.2419$.

6 Find the possible values of the angles A, B, C.....I in the range 0° to 360°. If you are using 3-figure tables, first correct each of the given decimals.

$\sin A = 0.6157$, $\sin B = -0.9563$, $\sin C = 0.9272$, $\cos D = -0.1736$, $\cos E = -0.7771$, $\cos F = 0.5299$, $\tan G = 3.7321$, $\tan H = 0.4040$, $\tan I = -0.9004$.

7 In the given right-angled triangle, $\sin A = \frac{4}{5}$. Write down $\cos B$.
What do you notice about $\sin A$ and $\cos B$?
What do you know about angle A and angle B?
Use your tables to find angles A and B.

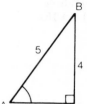

8 a) Given that $\sin 25° = 0.423$, write down $\cos 65°$.
b) Given that $\sin 72.5° = 0.954$, write down $\cos 17.5°$.
c) Given that $\cos 55.8° = 0.562$, write down $\sin 34.2°$.
d) If angle X + angle $Y = 90°$ and $\sin X = \frac{3}{7}$, what is $\cos Y$?

9 From the following list find the ratios which are equal in value to $\cos 40°$:
$\sin 50°$, $\cos 140°$, $\sin 140°$, $\sin 130°$, $\cos 220°$, $\sin 310°$, $\cos 320°$.

10 From the following list find those ratios which are equal in value to one another: $\sin 15°$, $\cos 15°$, $\sin 75°$, $\cos 75°$, $\sin 165°$, $\cos 165°$, $\sin 195°$, $\cos 345°$.

11 Taking values of x from 0 to 180 in intervals of 30, draw the graph of $y = \sin(x + 90)°$. Compare this curve with the graphs of a) $y = \sin x°$, b) $y = \cos x°$. What do you find?

12 If $\cos 40° = 0.766$, what is $\cos(-40°)$? Draw a sketch to illustrate your answer. Repeat question 11 for $y = \cos(x - 90)°$. What can you deduce from this? What transformation maps $y = \sin x°$ on to $y = \cos x°$?

13 Take values of x from 0° to 90° in intervals of 10° and draw the graph of $y = \sin 3x$.
On the same axes, over the same range of values for x, draw the graph of $y = 3 \sin x$. (Take x in intervals of 30°.)
The first of these curves represents an enlargement with scale factor $\frac{1}{3}$ of $y = \sin x$, in the direction of the x axis. Describe the enlargement in the second of these curves.

14 Draw the graph of $y = 2 + \cos x$ for $0° \leqslant x \leqslant 360°$.
For the same range of values of x, draw the graph of $y = \cos(x + 30°)$.
Compare both these curves with $y = \cos x$ and describe the transformations which have taken place in each case.

15 On the same axes draw the graphs of $y = 2 \sin x$ and $y = 2 - \sin x$. Take x from $0°$ to $270°$ in intervals of $30°$ for both. From your graph solve the equation $3 \sin x = 2$.

16 A mass is suspended by a spring which is hanging from a support. The mass is pulled down and let go. As it oscillates up and down its distance in cm from the support is given by $30 + 10 \cos (360t)°$ where t is the time in seconds.

a) Take t in intervals of a twelfth of a second and calculate the corresponding distances.
b) Draw a graph of distance against time covering a period of 1 second.
c) Considering the gradient of your graph, at what times is the mass *i*) at rest,
ii) moving upwards with maximum speed,
iii) moving downwards with maximum speed.
d) The figure shows three positions of the mass at different times. Copy the figure into your book and mark on the times at which the mass is in each of the three given positions.

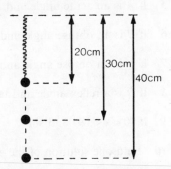

17 The formula $h = 11 \cdot 5 + 4 \cdot 5 \cos (30d)°$ gives the *approximate* length of time in hours between sunset and sunrise on the last day of each month during one year. Take values for d from 0 to 12 and draw a graph of h against d.

a) At what time of the year do you think this graph starts?
b) How many hours of daylight are there on the longest day?

18 The depth of water on a certain day alongside a quay is given by the formula $d = 5 - 2 \sin (30h)°$ where d is the depth in metres and h is the number of hours after midnight.
Draw a graph to show the depth of water for a twenty-four hour period from midnight.
A ship requires at least 4 m of water when arriving unloaded and at least 6 m to sail loaded. From your graph answer the following:

a) Between what times must the ship arrive?
b) Between what times later in the day can she sail?
c) What is the time interval between successive high tides?

*19 Question 18 gives a simple but inaccurate picture of the tides. A rather more accurate picture is obtained using the formula $d = 5 - 2 \sin (29h)°$.
Draw the new graph for $h = 0$ to 25. Using this revised formula, what is the time interval between two successive high tides?

4C Miscellaneous

1 From the following list find those ratios which are equal in value to one another:
$\sin 20°$, $\cos 20°$, $\sin 70°$, $\cos 70°$, $\sin 160°$, $\cos 160°$.

2 Solve the equation $\sin x° = \cos 35°$ for $0 \leqslant x \leqslant 180$.

3 Solve the equation $\cos x° = \sin 65°$ for $0 < x < 360$.

4 If $0° \leqslant x \leqslant 180°$, find possible solutions of the equation $1 = 2 \sin X$.

5 If A is an acute angle and $\cos A = \frac{4}{5}$, write down $\sin A$ and $\tan A$.

6 If B is an obtuse angle and $\sin B = \frac{4}{5}$, write down $\cos B$ and $\tan B$.

7 If C is an obtuse angle and $\cos C = -\frac{5}{13}$, write down $\sin C$ and $\tan C$.

8 If D is a reflex angle and $\tan D = \frac{5}{12}$, write down $\sin D$ and $\cos D$.

9 If $90 < x < 180$ and $\sin x° = \frac{12}{13}$, write down the value of $1 + \cos x°$.

10 Find one solution of the equation $\sin (4x + 10)° = 1$.

11 *The Inverse Notation*
The inverse notation $A = \sin^{-1}(0.50)$ reads 'A is the angle whose sine is 0·50'. It is sometimes useful to be able to express angles in this way. If all the angles in this question lie between $0°$ and $180°$, calculate the possible values of angles P, Q, R, S and T.

$$P = \sin^{-1}(\tfrac{2}{3}) \qquad Q = \sin^{-1}(\tfrac{2}{5}) \qquad R = \cos^{-1}(\tfrac{4}{7}) \qquad S = \tan^{-1}(\tfrac{3}{8}) \qquad T = \tan^{-1}(\tfrac{4}{3}).$$

12 If the angles $A, B, C \ldots F$ lie in the range $0°$ to $360°$, find all the possible values of $A \ldots F$.

$$A = \tan^{-1}(2\cdot9208) \qquad B = \tan^{-1}(-0\cdot3057) \qquad C = \sin^{-1}(-0\cdot5446)$$
$$D = \cos^{-1}(0\cdot2756) \qquad E = \cos^{-1}(-0\cdot9068) \qquad F = \sin^{-1}(-0\cdot8587)$$

13 Taking the smallest value between $0°$ and $360°$ for each of the given angles, and using your tables, find the value of:

a) $\sin^{-1}\frac{1}{2} + \cos^{-1}\frac{1}{2}$ b) $\sin^{-1} 1 + \cos^{-1} 0$ c) $\sin^{-1} 0\cdot3 + \sin^{-1} 0\cdot85$
d) $\sin^{-1} 0\cdot8 + \sin^{-1}(-0\cdot6)$ e) $\cos^{-1} 0\cdot3 + \cos^{-1}(-0\cdot8)$

Where appropriate give your answers to the nearest tenth of a degree.

14 State whether the following are true or false:

a) $\sin^{-1} 0\cdot866 + \cos^{-1} 0\cdot5 = \cos^{-1}(-0\cdot5)$
b) $\tan^{-1} 1 + \tan^{-1} 2 = \tan^{-1} 3$ c) $\sin^{-1} 0\cdot2 + \cos^{-1} 0\cdot2 = \tan^{-1} 2\cdot2$
d) $\sin^{-1} 0\cdot866 = 2 \sin^{-1} 0\cdot5$ e) $\cos^{-1} 0\cdot5 = \tan^{-1} 1\cdot732$

✱15 The unit square $OABC$ with vectors at $(0, 0)$, $(1, 0)$ etc. is sheared by the matrix $\begin{pmatrix} 1 & 0\cdot5 \\ 0 & 1 \end{pmatrix}$ to $OAB'C'$.

a) What are the co-ordinates of C'?
b) The angle COC' is called 'the angle of shear' and is equal to $\tan^{-1} \dfrac{CC'}{OC}$. Give its value to the nearest half degree.

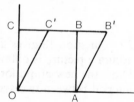

34

*16 State the angles of shear for the following matrices. Give the angle
i) in $\tan^{-1}$ form, *ii)* to the nearest degree, between $-90°$ and $90°$.

  a) $\begin{pmatrix} 1 & 2 \\ 0 & 1 \end{pmatrix}$ *b)* $\begin{pmatrix} 1 & 1 \\ 0 & 1 \end{pmatrix}$ *c)* $\begin{pmatrix} 1 & -1 \\ 0 & 1 \end{pmatrix}$ *d)* $\begin{pmatrix} 1 & -0\cdot6 \\ 0 & 1 \end{pmatrix}$ *e)* $\begin{pmatrix} 1 & k \\ 0 & 1 \end{pmatrix}$

17 Find the maximum value of y if $0° \leqslant x \leqslant 180°$ and $y = 3 \sin x$.

18 Find the maximum value of y if $0° \leqslant x \leqslant 180°$ and $y = 5 \sin 2x$. What are the values of x for which y is a maximum?

19 Find the minimum value of y if $0° \leqslant x \leqslant 180°$ and $y = 2 \cos 2x$. For what values of x is y a minimum?

20 *Identities*
For the right-angled triangle shown in the diagram, write down a relation between the lengths a, b and c. Divide both sides of your equation by b^2. Write down expressions for $\sin A$ and $\cos A$ in terms of a, b and c. By substituting for these ratios in your equation, find a relationship between $\sin^2 A$ and $\cos^2 A$.

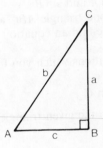

21 If $A = 30°$, use your tables to find the value of $\sin^2 A + \cos^2 A$.

22 Repeat question 21 for *a)* $A = 43°$ *b)* $A = 17°$ *c)* $A = 54°$ *d)* $A = 129°$
e) $A = 248°$.

23 If $0° \leqslant A \leqslant 360°$, use the relationship which you have found in question 20 to solve the equation $7 \sin^2 A - \cos^2 A = 1$.

24 Solve each of the following equations. Give all values of A for which $0° \leqslant A \leqslant 360°$.

  a) $3 \cos^2 A - \sin^2 A = 2$ *b)* $\cos^2 A - \cos A = 0$
  c) $3 \sin^2 A = 2 \sin A + 1$ *d)* $\sin^2 A - \sin A = \cos^2 A + 2$

25 Find the length of the hypotenuse in the right-angled triangle shown in the diagram. Hence write down the values of $\sin A$, $\cos A$ and $\tan A$. What is the ratio $\dfrac{\sin A}{\cos A}$?

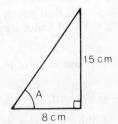

26 For any angle $\dfrac{\sin A}{\cos A} = \tan A$. Use this to find all the values of A if
$0° \leqslant A \leqslant 180°$ and:

  a) $\sin A = \cos A$ *b)* $3 \cos A - 2 \sin A = 0$
  c) $\sin A + \cos A = 0$ *d)* $5 \sin A + 3 \cos A = 0$.

5 More Trigonometry

5A The Sine Rule

1 Triangle *ABC* is any triangle, and the
perpendicular height through *C* meets *AB* at
D. The sides are lettered using the convention
that the side opposite *A* is *a*, the side opposite
B is *b*, etc. From triangle *ADC*, $CD = b \sin A$.
Write down a second expression for *CD* using
triangle *BCD* and hence find an equation
which relates sin *A* and sin *B*.

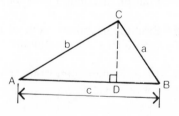

Start with a second triangle *ABC* and draw in the altitude through *B*. Using a similar
method, this time find an equation which relates sin *A* and sin *C*.

2 The two equations which you found in question 1 put together give the sine rule:
$$\frac{a}{\sin A} = \frac{b}{\sin B} = \frac{c}{\sin C}$$

Using this rule in the given triangle, calculate:

a) length *a*
b) angle *C*
c) length *c*.

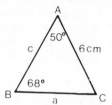

3 Repeat question 2 for triangle *ABC* in which angle $A = 57°$, angle $B = 42°$ and
length $b = 8.5$ cm.

4 The triangles in this question should not be drawn to scale.

a) Draw triangle *EFG* in which $e = 5$ cm, angle $E = 49°$, angle $F = 52°$. Calculate the
length *f*.
b) Draw triangle *LMN* in which $m = 10.4$ cm, angle $L = 38°$, angle $M = 69°$. Calculate
the length *l*.
c) Draw triangle *PQR* in which $q = 17$ cm, angle $P = 67°30'$, angle $R = 34°30'$.
Calculate angle *Q* and hence length *p*.
d) Draw triangle *XYZ* in which $x = 34$ cm, angle $X = 49°24'$, angle $Z = 61°12'$.
Calculate angle *Y* and hence length *y*.

5 In triangle *ABC* one form of the sine rule is $\dfrac{a}{\sin A} = \dfrac{b}{\sin B} = \dfrac{c}{\sin C}$.

Rearrange this in the form $\dfrac{\sin A}{a} = $.

Complete this to link the angles *A*, *B* and *C*. This is another form of the sine rule.

6 Using the version of the sine rule which
you have just found, calculate angle *A* in the
given triangle.

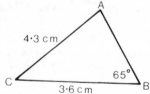

7 In triangle *ABC*, $b = 14$ cm, $c = 12.2$ cm and angle $B = 73°$.
Calculate a) angle *C*, b) angle *A*, c) length *a*.

8 For each part of this question draw a triangle (not to scale), letter the sides and angles appropriately and answer the questions using the sine rule.

a) Calculate angle F in triangle DEF where $d=7$ cm, angle $D=71°$, $f=6.3$ cm.
b) Calculate angle T in triangle RST where $s=15.3$ cm, $t=14.6$ cm, angle $S=59°$.
c) In triangle LMN, $l=21$ cm, $m=25$ cm, angle $M=79°$. Calculate angle L and hence find angle N.
d) In triangle PQR, $p=17.2$ cm, $r=11.7$ cm, angle $P=68°$. Calculate angle R and hence find angle Q.

9 In triangle XYZ, $x=4.2$ cm, $y=3.7$ cm, angle $X=59°$. Calculate

a) angle Y b) angle Z c) length z.

10 In triangle PQR, $q=19.4$ cm, angle $P=62°$, angle $R=65°30'$. Calculate

a) angle Q b) length p c) length r.

5B   Examples Containing Obtuse Angles

1 In triangle ABC, $a=9.4$ cm, $b=7.8$ cm, angle $A=120°$. Calculate angle B.

2 In triangle XYZ, $x=14.6$ cm, $z=17.5$ cm, angle $Z=135°$. Calculate angle X.

3 In triangle LMN, $l=8.9$ cm, $m=11.6$ cm, angle $M=142°$. Calculate
a) angle L b) angle N c) length n.

4 In triangle PQR, $q=62$ cm, $r=55$ cm, angle $Q=115°$. Calculate
a) angle R b) angle P c) length p.

5 In triangle ABC, angle $A=100°$, angle $B=35°$, $b=5.3$ cm. Calculate
a) length a b) angle C c) length c.

6 In triangle DEF, angle $F=39°12'$, angle $E=114°48'$, $d=10.3$ cm. Calculate
a) angle D b) length e c) length f.

7 In triangle XYZ, $y=4.8$ cm, $x=3.5$ cm, angle $X=42°$. Do you know whether angle Y is acute or obtuse? Calculate two possible values for angle Y. By drawing the triangle accurately, find whether both of your answers are possible.

8 Repeat question 7 for   a) $y=6.0$ cm, $x=5.0$ cm, angle $X=50°$ b) $y=4.2$ cm, $x=2.6$ cm, angle $X=30°$ c) $y=3.5$ cm, $x=2.8$ cm, angle $X=53.1°$ d) $y=4.4$ cm, $x=5.5$ cm, angle $X=64°$.

9 In the parallelogram $ABCD$, $AB=DC=5.3$ cm, $AD=BC=3.9$ cm, and the diagonal AC makes an angle of $43°$ with the side AB. Calculate the angles of the parallelogram.

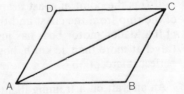

10 In parallelogram $PQRS$, PR is the longer diagonal and is of length 14.5 cm. $QR=PS=10.8$ cm, and the angle between PR and PQ is $32°$. Calculate the angles of the parallelogram and the lengths of the other two sides.

11 In parallelogram $WXYZ$, the shorter diagonal WY is 13·9 cm long. $WZ = XY = 9·2$ cm, and two of the angles of the parallelogram are 53°. Calculate the angles which WY makes with the sides, and find the lengths of the other pair of sides of the parallelogram.

Give your answers to numbers 12–16 correct to 2 significant figures.

12 The captain of a ship first sights a lighthouse on a bearing of 032° and after sailing due north for 8 km finds that the lighthouse is on a bearing of 079°. At that time how far is the ship from the lighthouse?

13 Two of the wires which are holding up a mast are fastened to the same point on the mast and to two points on the ground 28 m apart and in line with the bottom of the mast. Both points are on the same side of the mast. If one wire is inclined at 40° to the horizontal and the other at 65° to the horizontal, find the lengths of the two wires.

14 Two points X and Y are 35 m apart at the same level on opposite banks of a river. A ship is moored in the river between X and Y. The angle of elevation of the top of the mast of the ship from X is 12° and from Y the angle of elevation is 16°. Calculate the distance of X from the top of the mast and hence calculate the height of the mast above the level XY.

15 A man standing on level ground due north of a church tower finds that the angle of elevation of the top of the flag on the top of the tower is 54°. He walks 65 m due south and is then south of the foot of the tower. The angle of elevation of the top of the flag is now 48°. Find the direct distance of the top of the flag from the man in his final position and hence the height of the top of the flag above the ground. (Ignore the height of the man.)

16 A boy is walking along a straight, level section of the bank of a river from west to east. He observes that a tree on the opposite bank is on a bearing of 039° 30′. After he has walked 75 m the bearing of the same tree is 335° 30′. How far is he from the tree at that time and what was his shortest distance from the tree as he walked along the bank?

✱17 A stationary ship, suspected of carrying contraband, is sighted from two observation stations A and B 5 km apart on an approximately straight stretch of coast. The bearing of B from A is 303° and the bearing of the ship S from A is 318°. The bearing of S from B is 114°.

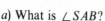

   a) What is $\angle SAB$?
   b) What is the bearing of A from B?
   c) What is $\angle SBA$?
   d) Using the sine rule, calculate the length of AS to 3 s.f.
   e) If D is the point of coast nearest to S, find the distance of the ship from the coast and the length of AD.
   f) If a loaded motor boat has just been launched from the ship and it is unlikely to travel at more than 15 km/h, how soon should the customs men reach D to make a perfect interception?

✱18 An aircraft on a training flight passes directly over an observation post P and flies on a level course which takes it directly over another observation post Q, 9·6 km distant from P and at the same height above sea level. At 1215 GMT the angle of elevation of the aircraft is 21° from P and 8° from Q. At 1216 GMT the angle of elevation from P is 6°. If A is the position of the aircraft at 1215 and B is its position at 1216 GMT,

a) find the length of PA in km.
b) find the height of the aircraft above PQ in metres.
c) find the length in km of PD and DE if D and E are the projections of A and B on PQ.
d) what is the speed of the aircraft in km/h?
e) what is the angle of elevation of the aircraft from B at 1216 GMT?

5C Further Examples Using the Sine Rule

1 In triangle ABC, $AB = 36.5$ cm, $BC = 28.9$ cm and angle $C = 37°$. Calculate angle A.

2 Given that angle $P = 42° 30'$, angle $Q = 65° 30'$ and $PR = 46.5$ cm, calculate the lengths of the other two sides of triangle PQR.

3 In triangle XYZ, $XY = 13$ cm, $YZ = 11.8$ cm and angle $Z = 82°$. Calculate the other two angles and the length ZX.

4 In a parallelogram the shorter sides are of length 5.8 cm, the diagonals bisect each other at 40° and the shorter diagonal measures 8.5 cm. Find the length of the other diagonal.

5 The diagram shows a symmetrical kite in which the angles at A and C are each 18° and the reflex angle at D is 226°. If $BD = 20$ cm, find the lengths of the sides of the kite.

6 The diagram shows a trapezium $PQRS$ in which angle $Q = 90°$. If $PQ = 12$ cm, $QR = 7$ cm and $PS = 8.5$ cm, calculate

a) angle QPR *b)* PR
c) angle PSR *d)* angle QPS
e) SR.

7 The steps (AB) of a children's slide are inclined at 40° to the level ground and the slide itself (AC) is inclined at 25° to the ground. If the distance between the foot of the steps and the foot of the slide (taken as C) is 10 m, calculate the length of the slide. (*Note* In actual practice the slide is 'flattened' at the bottom as shown.)

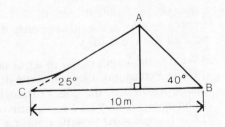

8 From a port P the bearing of island I is 050° and the bearing of island J is 125°. The bearing of J from I is 165°. A ship leaves P and sails the 38 km to J. Calculate the distance it has to sail to return to P via I.

9 Radius of Circumcircle

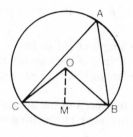

The diagram shows triangle ABC circumscribed by a circle, centre O. M is the foot of the perpendicular from O to CB. If angle CAB is $40°$ and $CB=6$ cm, write down

  a) angle COB b) angle COM c) CM
  d) the radius of the circle, i.e. CO.

10 Repeat question 9 using 'A' for angle CAB and 'a' for CB, and hence write an equation which begins $2R=$, where R is the radius of the circumcircle. Using the sine rule, write two other expressions for $2R$.

11 Using the formula which you have just found, calculate the radius of the circumcircle of each of these triangles:

  a) triangle ABC in which $AC=9.4$ cm, angle $B=72°$.
  b) triangle PQR in which $QR=21$ cm, angle $P=48°$.
  c) triangle DEF in which $EF=14.7$ cm and $D=35°$.

12 If in triangle XYZ angle X is an obtuse angle and $YZ=x$, prove that R the radius of the circumcircle is still given by the formula $2R=\dfrac{x}{\sin X}$.

(*Hint* Complete a cyclic quadrilateral $PZXY$.)

13 Find the radius of the circumcircle of triangle RST in which angle $R=138°$ and $ST=7.5$ cm.

14 Calculate the angles which a chord of length 12 cm drawn in a circle of radius 8·2 cm subtends at the circumference of the circle.

15 A chord of length 11·9 cm is drawn in a circle of radius 17·3 cm. Calculate the possible angles which the chord can subtend at the circumference of the circle.

5D The Cosine Rule

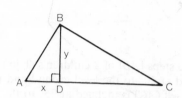

1 Triangle ABC is any triangle and the altitude through B meets AC at D. Use the usual convention for the length of the sides of a triangle, i.e. denote the length of the side opposite A by a, the side opposite B by b, etc. Call AD x and BD y.

  a) What is the length of CD in terms of x?
  b) Using Pythagoras' theorem, write down a relationship between i) the sides of triangle ABD ii) the sides of triangle BCD.
  c) Eliminate y from your two equations and get an equation in x.
  d) Use triangle ABD to write down an expression for x in terms of the cosine of A.
  e) Substitute this expression for x in your single equation and hence prove that $a^2=b^2+c^2-2bc\cos A$. This is the Cosine Rule.

2 Use the above formula to find the length a in each of the triangles ABC:

a) $b=5$ cm, $c=2$ cm, angle $A=72°$. c) $b=11$ cm, $c=8$ cm, angle $A=37°$.
b) $b=4$ cm, $c=7$ cm, angle $A=60°$. d) $b=3·5$ cm, $c=5·2$ cm, angle $A=52°$.

3 The cosine rule can also be written in the form:

$b^2 = \ldots\ldots$ (the expression on the right including $\cos B$) and
$c^2 = \ldots\ldots$ (the expression on the right including $\cos C$).
Write down these two forms of the cosine rule in full.
Use the appropriate version of the rule in each of the following triangles.

a) Calculate length c of triangle ABC in which $a=3$ cm, $b=5$ cm, angle $C=25°$.
b) Calculate length b of triangle ABC in which $a=22$ cm, $c=27$ cm, angle $B=49°$.
c) Calculate length b of triangle ABC in which $a=4·3$ cm, $c=6·5$ cm, angle $B=31° 30'$.
d) Calculate length c of triangle ABC in which $a=16$ cm, $b=11$ cm, angle $C=82° 15'$.

4 Use the cosine rule in each of the following triangles.

a) Calculate length g of triangle EFG in which $e=7$ cm, $f=9$ cm, angle $G=38°$.
b) Calculate length m of triangle LMN in which $l=15$ cm, $n=12·5$ cm, angle $M=42° 45'$.
c) Calculate length p of triangle PQR in which $q=3·6$ cm, $r=5·2$ cm, angle $P=25° 30'$.
d) Calculate length x of triangle XYZ in which $y=31$ cm, $z=28$ cm, angle $X=74° 30'$.

5 Rearrange the formula $a^2 = b^2 + c^2 - 2bc \cos A$ to make $\cos A$ the subject.

6 Find angle A of triangle ABC in which $a=6$ cm, $b=5$ cm and $c=7$ cm.

7 Using another form of the cosine rule, make $\cos B$ the subject. Find angle B of triangle ABC in which $a=5·3$ cm, $b=6·0$ cm and $c=4·8$ cm.

8 Calculate angle C of triangle ABC in which $a=22$ cm, $b=29$ cm and $c=19$ cm.

9 Calculate angle P of triangle PQR in which $p=6·7$ cm, $q=5·5$ cm and $r=4·9$ cm.

10 In triangle XYZ, $x=6$ cm, $y=8$ cm, $z=11$ cm. Calculate
a) angle X by the cosine rule   b) angle Y by the sine rule.

11 In triangle LMN, $l=13$ cm, $n=9·5$ cm, angle $M=36° 15'$. Calculate

a) length m by the cosine rule   b) angle N by the sine rule.

12 In triangle EFG, $f=31$ cm, $g=43$ cm, angle $E=65° 42'$. Calculate

a) length e by the cosine rule   b) angle F by the sine rule.

5E Cosine Rule: examples involving obtuse angles

1 Use the cosine rule in each of the following.

a) Calculate length c of triangle ABC in which $a=3·5$ cm, $b=5·8$ cm, angle $C=120°$.
b) Calculate length l of triangle LMN in which $m=14$ cm, $n=12·8$ cm, angle $L=134°$.
c) Calculate length q of triangle PQR in which $p=35$ cm, $r=27$ cm, angle $Q=158° 30'$.

d) Calculate length x of triangle XYZ in which $y=19\cdot5$ cm, $z=21\cdot3$ cm, angle $X=114°\,15'$.

2   In each of the following triangles calculate the size of the largest angle using the cosine rule.

  a) Triangle RST in which $r=5$ cm, $s=7$ cm and $t=10$ cm.
  b) Triangle DEF in which $d=29$ cm, $e=21$ cm, $f=11$ cm.
  c) Triangle JKL in which $j=8\cdot2$ cm, $k=14\cdot3$ cm, $l=7\cdot1$ cm.
  d) Triangle PQR in which $p=32$ cm, $q=25$ cm, $r=12$ cm.

3   In triangle ABC, $AB=12\cdot4$ cm, $BC=8\cdot9$ cm, angle $B=125°$. Calculate

  a) length CA using the cosine rule.
  b) the smallest angle of the triangle using the sine rule.

4   In triangle XYZ, $XY=4\cdot9$ cm, $XZ=9\cdot3$ cm, angle $X=143°$. Calculate

  a) length YZ *b*) angle Z.

5   In triangle PQR, $PQ=15\cdot8$ cm, $QR=10\cdot2$ cm, and $RP=7\cdot6$ cm. Calculate

  a) angle R using the cosine rule   *b*) angle P using the sine rule.

6   Find the lengths of the diagonals of a parallelogram in which the acute angles measure $51°$ and the sides $4\cdot8$ cm and $7\cdot5$ cm.

7 $ABCD$ is a parallelogram in which $AB=CD=14$ cm, $BC=AD=6\cdot9$ cm and angle $ABC=68°$. Calculate

  a) angle BCD *b*) AC *c*) BD.

8   The diagonals PR and QS of parallelogram $PQRS$ bisect each other at $45°$. Find the lengths of PQ and QR if $PR=22$ cm and $QS=16\cdot4$ cm.

9   A ship sails from port on a bearing of $130°$. After 15 km the course is changed to $075°$. How far from the port will the ship be after sailing another 22 km?

10   Three buoys A, B and C are used to mark the approach to a harbour. B is due east of A and 180 m away. C is 90 m from A and 120 m from B, and to the north of AB. Find the bearings of C from A and from B. (Give your answers to the nearest degree.)

11   In triangle PQR, $PQ=8$ cm, $QR=6$ cm and $RP=5$ cm. QR is produced to S so that $RS=3\cdot5$ cm. Calculate

  a) angle PRQ *b*) length PS.

12   In triangle WXY, $WX=12\cdot3$ cm, $XY=14$ cm and $YW=9\cdot8$ cm. XY is produced to Z so that $YZ=7$ cm. Calculate the length WZ, by first finding the cosine of angle XYW.

13   Two straight roads meet at an angle of $50°$. The King's Arms is 500 m from the crossroads on one road and a filling station is 700 m from the crossroads on the other road. How far are they apart, as the crow flies? (Two possible answers. Give each to the nearest 10 metres.)

14

(i)

(ii)

Fig (i) shows a glider being towed by a tractor using a cable of length 80 metres. When the tractor has travelled 100 metres, the glider is airborne and the cable makes an angle of 30° with the ground.

Fig. (ii) shows the conditions diagrammatically, the tractor and glider being originally at A and B and finally at C and D. Calculate the approximate distance the glider has travelled, i.e. the length BD.

(Its actual path will, of course, be a curve.)

15· An ellipse is drawn by using a loop of thread 14 cm long and placing it round two pins A and B 4 cm apart. The point of a pencil is inserted in the loop and the loop pulled tight. The ellipse is drawn by moving the pencil in the direction shown. AB is produced both ways to cut the completed ellipse in P and Q. RS is perpendicular to PQ and goes through O, the mid-point of AB.

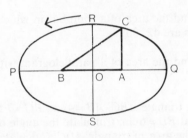

a) What is the angle CAQ when i) $CA = 4.5$ cm   ii) $CA = 3.8$ cm?
Give your answer to the nearest tenth of a degree.
b) What is the length of the major axis, PQ?
c) What is the length of the minor axis, RS?

16 Amateur groundsmen sometimes lay out a right angle by using a loop of thin cord knotted at A, B and C as shown in the diagram. If knot B is wrongly positioned so that $BC = 3.9$ m and $BA = 3.1$ m, what angle would be laid out at B?

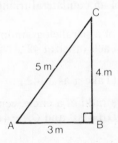

5F   The Area of a Triangle

1   In triangle ABC, the altitude through A meets BC at D. Given that $AC = 6$ cm, $BC = 9$ cm and angle $C = 30°$, calculate

a) the perpendicular height AD
b) the area of the triangle ABC.

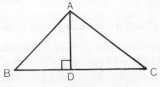

2   Using a, b and c for the lengths of the sides of triangle ABC and the diagram in question 1, write down

  a) AD by considering triangle ABD and angle B
  b) the area of triangle ABC
  c) AD by considering triangle ACD and angle C
  d) the area of triangle ABC.

You have now found two forms of the formula for finding the area of a triangle ABC. Write down a third one.

3   By using the formula $\frac{1}{2}ab\sin C$, or a similar formula, find the areas of each of the following triangles.

  a) Triangle ABC in which angle $A=40°$, $b=7$ cm, $c=10$ cm.
  b) Triangle LMN in which angle $N=72°$, $m=16$ cm, $l=11\cdot5$ cm.
  c) Triangle PQR in which $p=3\cdot8$ cm, $q=5\cdot3$ cm, angle $R=51°$.
  d) Triangle XYZ in which $x=21$ cm, $z=27$ cm, angle $Y=35°$.

4   Find the area of triangle ABC in which $AB=AC=7$ cm and angle $A=110°$.

5   Find the area of an isosceles triangle in which the equal sides are each $17\cdot5$ cm and the vertical angle is $68°\,30'$.

6   Find the area of a rhombus in which the sides are of length 10 cm and the acute angles are $42°\,30'$.

7   Find the area of a parallelogram with sides of $4\cdot7$ cm and $8\cdot3$ cm and acute angles of $57°$.

8   In triangle ABC, $AB=4$ cm, $BC=7\cdot5$ cm and angle $B=80°$. AB is produced to D so that $BD=6$ cm. Calculate the angle of triangle ABC and of triangle CBD and hence find the area of triangle ADC. Calculate the total area by a different method. Which method is the simpler in this case? Discuss.

9   Find the area of an equilateral triangle of side 8 cm.

10   Find the area of a parallelogram in which the diagonals are of length $9\cdot8$ cm and $14\cdot5$ cm and bisect each other at $42°$.

Segments of circles (Take π as $3\cdot142$.)

11 OA and OB are radii of a circle, centre O. Given that angle $AOB=90°$ and $OA=4$ cm, calculate

  a) the area of the sector AOB
  b) the area of the triangle AOB
  c) the area of the minor segment cut off by AB.

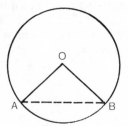

12   Using the diagram above, if $OA=12$ cm and angle $AOB=60°$, repeat question 11.

13   Repeat question 11 in each of the following cases:

  a) diameter of circle $=10$ cm, angle of sector $=36°$
  b) radius of circle $=18$ cm, angle of sector $=150°$
  c) diameter of circle $=24$ cm, angle of sector $=45°$.

14 A chord of length 10 cm is drawn in a circle of radius 9 cm. Find the area of the minor segment cut off by the chord.

15 A chord of length 10 cm is drawn in a circle of radius 13 cm. Calculate the ratio of the areas of the major and minor segments into which the chord divides the circle. Give your answer in the form $n:1$.

16 *The 's' Formula for the Area of a Triangle*

The area of a triangle is given by the formula $A = \sqrt{s(s-a)(s-b)(s-c)}$ where $2s = a+b+c$.
In triangle ABC, if $a = 5$ cm, $b = 8$ cm, $c = 9$ cm, find the area using the given formula.

17 Find the area of triangle PQR in which $p = 6$ cm, $q = 3$ cm, $r = 7$ cm.

18 Find the area of triangle XYZ in which $XY = XZ = 15$ cm and $YZ = 22$ cm.

19 Use the 's' formula to find the area of an equilateral triangle of side 12 cm.

20 The diagram shows a pentagon in which angle D = angle C = 90°, $AB = AE = 11$ cm, $BC = ED = 9$ cm and $CD = 16$ cm. Calculate

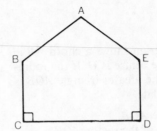

  a) the area $BCDE$
  b) the area ABE
  c) the total area of the pentagon.

5G Miscellaneous

1 In triangle ABC, $a = 2$ cm, $b = 3$ cm, $c = 4$ cm. Find cosine A, leaving your answer in fraction form.

2 Find the cosine of the largest angle in a triangle with sides of 6 cm, 11 cm and 16 cm. Leave your answer as a fraction.

3 In triangle DEF, $f = 3$ cm, angle $E = 60°$ and angle $F = 40°$. Write an expression for DF in terms of the sines of the angles.

4 In triangle PQR, $q = 5$ cm, angle $Q = 85°$ and angle $R = 30°$. Write an expression for QR in terms of the sines of the angles.

5 In triangle XYZ, $x = 11$ cm, $y = 6$ cm and $z = 9$ cm. Write down the value of $\sin Y / \sin Z$.

6 In triangle LMN, $l = 5$ cm, $n = 8$ cm and angle $M = 30°$. Calculate the area of the triangle.

7 The area of triangle RST is 24 cm². If $r = 14$ cm and $s = 15$ cm, find the value of $\sin T$. Leave your answer as a fraction.

8 In triangle ABC, $a = 6$ cm, $c = 10$ cm and $\cos B = \frac{2}{15}$; find the value of b^2.

9 In triangle XYZ, $y = 8$ cm, $\sin X = \frac{1}{6}$ and $\sin Y = \frac{2}{3}$. Calculate x.

10 In triangle PQR, $PQ=5$cm, $QR=8$cm and $\cos Q=-\frac{11}{16}$. Calculate RP.

11 a) In triangle ABC, $a=11$cm, $b=9$cm, $c=15$cm. Calculate angles A and B.
b) In triangle XYZ, angle $X=68°$, $x=5\cdot5$cm, $z=4\cdot9$cm. Calculate angle Z.

12 Calculate the length of AC in triangle ABC where $BC=3$cm, $AB=7$cm and angle $C=70°$.

13 In triangle XYZ, $XY=2\cdot5$cm, $XZ=6\cdot5$cm and angle $X=32°$. Calculate the length YZ. If XZ is produced to P such that $ZP=3\cdot5$cm, calculate the length YP.

14 The diagram shows a kite $PQRS$ in which PQS is an equilateral triangle of side 6cm. If $RS=RQ=10$cm, calculate the area of the kite.

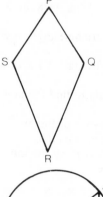

15 The diagram shows a chord AB in a circle whose centre is O. If the radius is 9cm and the area of triangle AOB is 20cm², calculate

a) the angle AOB
b) the area of the sector AOB
c) the area of the minor segment cut off by AB.

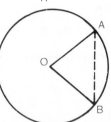

16 $ABCD$ is a parallelogram in which $AD=5$cm and angle $A=72°$. X and Y are points on AD and DC respectively such that BX and BY trisect angle B. Calculate
a) angle CYB b) the distance of Y from C.

17 $PQRS$ is a parallelogram in which angle $P=68°$, $PQ=8$cm and PR makes an angle of 28° with PQ. Calculate the length of the diagonal PR. If the bisector of angle S meets PR at A, find the distance of A from P.

18 Two boats leave a harbour at the same time. One travels at 6km/h on a bearing of 295° and the other on a bearing of 015° at 9km/h. Calculate the distance between the boats one hour after leaving the harbour, and the bearing of the second boat from the first. (Give your answer to the nearest degree.)

19 Two coastal stations A and B are 20km apart and B is on a bearing of 060° from A. A ship is observed at P on a bearing of 020° from A and 290° from B. Calculate the distance PB.
One hour later the ship is at Q which is on a bearing of 045° from A and due north of B. Calculate the speed of the ship, assuming that it has followed a straight course at a constant speed.

20 A plot of land is in the shape of a quadrilateral $ABCD$ in which $AB=35$m, $BC=42$m, $CD=30$m, angle $ABC=110°$ and angle $BCD=72°$. Calculate

a) AC b) angle ACB c) AD d) the total area.

5H Examples in 3 Dimensions including Plan and Elevation

Give your answers correct to 2 significant figures or the nearest degree.

1 The diagram shows a pyramid on a
horizontal rectangular base *ABCD*.
$AB=8$ cm, $AD=6$ cm, the vertex *V* is
vertically above *A* and $AV=9$ cm.

Calculate   *a*) *VC*   *b*) angle *ACV*.

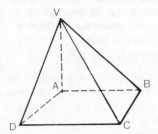

2 *VABCD* is a pyramid on a rectangular base *ABCD*. $AB=10$ cm, $BC=14$ cm, the
vertex *V* is vertically above *O* the centre of the rectangle, and $OV=7$ cm. Calculate

   a) the length of *OA*
   b) the length of the slant edges
   c) the length of *OE* where *E* is the mid-point of *AB*
   d) the inclination of *VAB* to the base
   e) the inclination of *VBC* to the base
   f) the inclination of the slant edges to the base.

3 *VPQRS* is a pyramid on a square base *PQRS* of side 12 cm. *V* is vertically above *O*,
the centre of the square base, and $OV=8$ cm. Calculate

   a) the length of the slant edges
   b) the angle each edge makes with the base.

If part of the top of the pyramid is cut away so that the vertex is now at *X*, the mid-
point of *VP*, find either by calculation or by scale-drawing

   c) the length *XR*
   d) the length *XQ*
   e) the angle that each of these slant edges makes with the base *PQRS*.

4 *VXYZ* is a pyramid on a triangular base. $XZ=ZY=13$ cm and $XY=10$ cm.
M is the mid-point of *XY* and *O* is the point in *MZ* such that $MO:OZ=1:2$. *V* is
vertically above *O*. If the inclination of *VZ* to *XYZ* is 45°, calculate

   a) the length of each of the slant edges
   b) the inclination of *VX* and *VY* to the horizontal
   c) the inclination of *VXY* to *XYZ*.

5 The diagram shows the vertical face of a solid which is on a horizontal rectangular
base *PQRS* where $PQ=12$ cm and $PS=15$ cm. The top of the solid is a horizontal
rectangle *ABCD* in which $AB=4$ cm and $BC=9$ cm. The perpendicular height of the
solid is 6 cm.

Use a suitable scale and draw

   a) the plan of the solid
   b) the elevation on a vertical plane parallel to *PQ*
   c) the elevation on a vertical plane parallel to *PS*

By further drawing where necessary find

   d) the inclination of each slanting face to the base
   e) the inclination of the edges *DS* and *CR* to the base.

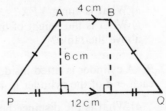

6 The diagram shows a wedge whose base *ABCD* is a horizontal rectangle. *ABFE* is a rectangular face in which *AB* = 6 cm and *BF* = 8 cm. This face is inclined at 30° to the base and the rectangular face *CDEF* is inclined at 45° to the base.

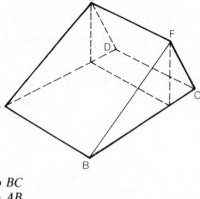

a) Calculate the length *BC*.

Part of the wedge is cut away so that the remainder forms a pyramid on *ABCD* as base with its vertex *V* at the mid-point of *EF*. Draw accurately

b) the plan of the pyramid
c) the elevation on a vertical plane parallel to *BC*
d) the elevation on a vertical plane parallel to *AB*.

By further drawing or by calculation find

e) the lengths of each of the slant edges of the pyramid
f) the inclination of each of these edges to the base.

7 A regular octahedron *ABCDEF* has edges of length 6 cm.

a) What is the angle between *AB* and *BE*?
b) What is the length of *BO* where *O* is the centre of *BCDE*?
c) What is the angle between *AB* and *BF*?
d) If *X* and *Y* are the mid-points of *BE* and *CD* respectively, *i)* state the length of *AX*, *ii)* name the angle between the faces *ABE* and *FBE*, *iii)* state its size in degrees.
e) What is *i)* the area of one face
ii) the total surface area of the octahedron?
f) What is the length of *AO*?
g) What is the volume of the octahedron?
h) What is the distance between opposite vertices?
i) Draw the plan of the octahedron (looking down on *A*) and also the elevation parallel to plane *ABFD*.

∗8 Face *BEF* of the octahedron in question 7 is laid flat on a horizontal table.

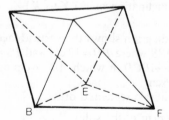

a) Name another face that is horizontal.
b) Draw a sketch (not to scale) of the quadrilateral *FXAY*. What kind of quadrilateral is it?
c) State the lengths of *i)* *AF*, *ii)* *XY*.
d) What is the angle between *AF* and *XY*?
e) What is angle *YFX*?
f) What is the height of the face *ACD* above the table?

9 A cylinder is fitted inside a hollow sphere of internal radius 5 cm. The radius of each end of the cylinder is 4 cm.

a) What is the distance between *O* the centre of the sphere and one end of the cylinder?

b) What is the height of the cylinder?
c) What is *i)* the total surface area of the cylinder, *ii)* its volume?
d) If P is a point on the circumference of one end of the cylinder, what angle does OP make with the axis of the cylinder?
e) If Q is a point on the circumference of the other end of the cylinder, what is *i)* the smallest possible value of angle POQ, *ii)* the largest possible value of angle POQ?

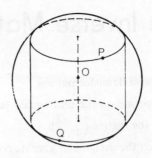

10 $ABCDEFGH$ is a cube of side $5\,cm$. P is a point on DB such that $DP = 2\,cm$.

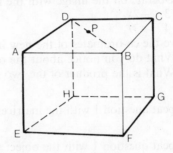

a) What is the length of DB?
b) What is the length of HP?
c) What angle does HP make with HD?
d) *i)* Name the angle between HP and the plane $EFGH$, *ii)* calculate its size.
e) Draw a rough sketch of triangle BPH. Mark on the lengths of the sides and calculate angle BHP.

∗11 Repeat question 10 for the rectangular prism in which $BC = 3\,cm$, $AE = 4\,cm$ and $DC = 5\,cm$.

∗∗12 In the rectangular prism in question 11 find the angle between the skew lines DF and AC. (*Hint* Draw triangle DBF. If J is the mid-point of DB, translate DF parallel to itself to go through J and cut BF at K. Solve triangle CJK.)

13 $ABCD$ is a regular tetrahedron of side $4\,cm$. O is the projection of A on the base BCD.

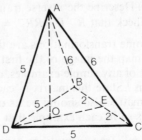

a) Draw a sketch of triangle BCD and mark on the position of O.
b) Describe the position of O in words.
c) Calculate the length of DO.
d) Calculate the length of AO.
e) If E is the mid-point of BC, what is the length of AE?
f) What is the area of triangle ABC?
g) What is the total surface area of the tetrahedron?
h) What is the angle between the edge AC and the base? What is the angle between the face ABC and the base?
i) Draw a plan and a front elevation of the tetrahedron.

∗14 In the tetrahedron $ABCD$, $AB = AC = 6\,cm$, $DB = DC = 5\,cm$, $BC = 4\,cm$ and $DA = 5\,cm$.

a) If E is the mid-point of BC, find the lengths of *i)* AE, *ii)* DE.
b) Using the cosine rule in triangle AED, calculate the angle between the face ABC and the base DBC.
c) If O is the projection of A on the plane BCD, find *i)* the position of O, *ii)* the length of AO.
d) Find the angle that AC makes with the base BCD, i.e. the angle ACO.
e) Draw a plan of the tetrahedron and a front elevation in a plane parallel to DC.

6 The Inverse Matrix and Determinants

6A Inverse Transformations

1 *a*) Operate on the unit square (shown) with the matrix $\begin{pmatrix} 3 & 2 \\ 4 & 3 \end{pmatrix}$.

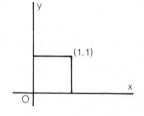

Sketch the image and give its co-ordinates.
b) Operate on the image with the matrix $\begin{pmatrix} 3 & -2 \\ -4 & 3 \end{pmatrix}$.

Give the co-ordinates of the new image.
c) What do you notice about the object in *a* and the final image in *b*?
d) What is the product of the two matrices in *a* and *b*?

2 Repeat question 1 with the matrices *a*) $\begin{pmatrix} 4 & 1 \\ 7 & 2 \end{pmatrix}$ *b*) $\begin{pmatrix} 2 & -1 \\ -7 & 4 \end{pmatrix}$.

3 Repeat question 1 with the object whose co-ordinates are (1, 1), (4, 1) and (4, 3) using the matrices *a*) $\begin{pmatrix} 3 & 1 \\ 2 & 1 \end{pmatrix}$ *b*) $\begin{pmatrix} 1 & -1 \\ -2 & 3 \end{pmatrix}$.

4 If you answered questions 1 to 3 correctly, you will know that in each case the second transformation mapped the image of the original object back on to the object. The second transformation is the *inverse* of the first transformation and the second matrix is the *inverse* of the first matrix. Now answer the following questions:

a) What transformation is defined by the matrix N, $\begin{pmatrix} \frac{1}{2} & 0 \\ 0 & \frac{1}{2} \end{pmatrix}$?

b) Describe geometrically the inverse transformation and write down its matrix.
c) This second matrix is the inverse of the first matrix. It can be called N^{-1}. Check that the two multiplied together in either order give the identity $\begin{pmatrix} 1 & 0 \\ 0 & 1 \end{pmatrix}$, i.e. that $N^{-1}N = NN^{-1} = I$.

5 *a*) What transformation is defined by the matrix P, $\begin{pmatrix} 1 & \frac{1}{4} \\ 0 & 1 \end{pmatrix}$?

b) Describe the inverse transformation geometrically and give its matrix. This matrix is P^{-1}, the inverse of P. Check that $P^{-1}P = PP^{-1} = I$.

6 *a*) What transformation is defined by the matrix R, $\begin{pmatrix} 0 & -1 \\ 1 & 0 \end{pmatrix}$?

b) Describe the inverse transformation geometrically. What is its matrix? Check that $R^{-1}R = RR^{-1} = I$.

7 Some transformations are their own inverses, i.e. when applied twice in succession they map the image of the first transformation back on to the original object. Can you think of any simple examples, choosing only those that can be represented by a (2×2) matrix? State this matrix in each case.
The matrix also should be its own inverse, i.e. MM (or M^2) should be I. Check that this is so in each example that you find.

✱*8* Some transformations cannot be represented by (2×2) matrices. A translation is a case in point. It can however be represented as a (3×3) matrix, and the image of this will map the image back on to the original object.

a) Write the position vector of the point (2, 4) as $\begin{pmatrix} 2 \\ 4 \\ 1 \end{pmatrix}$ and operate on it with

the matrix $\begin{pmatrix} 1 & 0 & 3 \\ 0 & 1 & 5 \\ 0 & 0 & 1 \end{pmatrix}$. What translation does this give?

b) Now operate on the image with the matrix $\begin{pmatrix} 1 & 0 & -3 \\ 0 & 1 & -5 \\ 0 & 0 & 1 \end{pmatrix}$.

You should get the original point (Ignore the one in the bottom row.)

c) Check that the product of the two matrices is I, i.e. that the second matrix is the inverse of the first.

d) Repeat a) and b) for the translation $\begin{pmatrix} 1 \\ 3 \end{pmatrix}$ starting with the point (3, 2).

6B Finding the Inverse Matrix

In 6A we saw that when a matrix A transforms a geometrical figure, its inverse A^{-1} 'undoes' the transformation and gives back the original figure, so that the two together are equivalent to the identity operation I, or $A^{-1}A = AA^{-1} = I$.

In 6B we examine the following problem. Given a matrix such as $\begin{pmatrix} 6 & 2 \\ 4 & 1 \end{pmatrix}$

how can we write down its inverse $\begin{pmatrix} -0.5 & 1 \\ 2 & -3 \end{pmatrix}$?

1 Multiply the following pairs of matrices in the order given:

a) i) $\begin{pmatrix} 2 & 1 \\ 5 & 3 \end{pmatrix}\begin{pmatrix} 3 & -1 \\ -5 & 2 \end{pmatrix}$ ii) $\begin{pmatrix} 3 & -1 \\ -5 & 2 \end{pmatrix}\begin{pmatrix} 2 & 1 \\ 5 & 3 \end{pmatrix}$

b) i) $\begin{pmatrix} 5 & 3 \\ 3 & 2 \end{pmatrix}\begin{pmatrix} 2 & -3 \\ -3 & 5 \end{pmatrix}$ ii) $\begin{pmatrix} 2 & -3 \\ -3 & 5 \end{pmatrix}\begin{pmatrix} 5 & 3 \\ 3 & 2 \end{pmatrix}$

c) i) $\begin{pmatrix} 8 & 5 \\ 3 & 2 \end{pmatrix}\begin{pmatrix} 2 & -5 \\ -3 & 8 \end{pmatrix}$ ii) $\begin{pmatrix} 2 & -5 \\ -3 & 8 \end{pmatrix}\begin{pmatrix} 8 & 5 \\ 3 & 2 \end{pmatrix}$

Write down the value of the determinant of the first matrix in each of a) i), b) i) and c) i). What do you notice about all six matrix products? Can you see how the second matrix in each of a) i), b) i) and c) i) was obtained from the first?

(*Hint* Calling the matrix $\begin{pmatrix} a & b \\ c & d \end{pmatrix}$ look first at a and d and then at b and c.)

2 Multiply the following pairs of matrices in the order given:

a) i) $\begin{pmatrix} 2 & 3 \\ 1 & 4 \end{pmatrix}\begin{pmatrix} 4 & -3 \\ -1 & 2 \end{pmatrix}$ ii) $\begin{pmatrix} 4 & -3 \\ -1 & 2 \end{pmatrix}\begin{pmatrix} 2 & 3 \\ 1 & 4 \end{pmatrix}$

b) i) $\begin{pmatrix} 3 & 2 \\ 6 & 5 \end{pmatrix}\begin{pmatrix} 5 & -2 \\ -6 & 3 \end{pmatrix}$ ii) $\begin{pmatrix} 5 & -2 \\ -6 & 3 \end{pmatrix}\begin{pmatrix} 3 & 2 \\ 6 & 5 \end{pmatrix}$

c) i) $\begin{pmatrix} 4 & 3 \\ 3 & 2 \end{pmatrix}\begin{pmatrix} 2 & -3 \\ -3 & 4 \end{pmatrix}$ ii) $\begin{pmatrix} 2 & -3 \\ -3 & 4 \end{pmatrix}\begin{pmatrix} 4 & 3 \\ 3 & 2 \end{pmatrix}$

Write down the value of the determinant of the first matrix in a) i). Divide the matrix products in a) by this determinant. Repeat with b) and c). What are your six final answers? Can you see how the second matrix in each of a) i), b) i) and c) i) was obtained from the first?

3 In question 1 the second matrix of each pair is the 'inverse' of the first matrix. Their product gives I.

In question 2, the product of the matrices gives $\begin{pmatrix} \Delta & 0 \\ 0 & \Delta \end{pmatrix}$ not $\begin{pmatrix} 1 & 0 \\ 0 & 1 \end{pmatrix}$, Δ being the determinant of the matrix.

What would you have to do to the second matrix of each pair so that it becomes the inverse of the first matrix? Do this to one or more of the matrices and check that the product of the pair is now I.

4 Now try a different method. Find the inverse of the matrix $\begin{pmatrix} 5 & 1 \\ 9 & 2 \end{pmatrix}$ algebraically.

Calling the inverse $\begin{pmatrix} a & b \\ c & d \end{pmatrix}$, then $\begin{pmatrix} 5 & 1 \\ 9 & 2 \end{pmatrix}\begin{pmatrix} a & b \\ c & d \end{pmatrix} = \begin{pmatrix} 1 & 0 \\ 0 & 1 \end{pmatrix}$.

Now find the product of the two matrices on the left side of this equation.

It starts $\begin{pmatrix} 5a+c & \\ & \end{pmatrix}$. Then equate each of the four terms of this product matrix to the corresponding term of the identity matrix $\begin{pmatrix} 1 & 0 \\ 0 & 1 \end{pmatrix}$.

This will give you four equations, the first being $5a+c=1$. What are the others?

Two of these equations are in a and d only. Solve them and find a and d. The other two are in b and c only. Solve them and find b and c. You can now write down the inverse matrix. What is it?

Check that the product of the original matrix and the inverse you have found is I.

5 Use the same procedure as in question 4 to find the inverse of $\begin{pmatrix} 3 & -1 \\ 8 & -2 \end{pmatrix}$.

What is the determinant of this matrix?

6 You should now be able to see that to find the inverse of the matrix $\begin{pmatrix} a & b \\ c & d \end{pmatrix}$,

i) change over a and d, ii) change the signs of b and c, iii) divide every term in the matrix so formed by the determinant of the original matrix.

Thus for the matrix $\begin{pmatrix} 3 & 2 \\ -4 & -2 \end{pmatrix}$ the determinant is 2, so the inverse is

$\frac{1}{2}\begin{pmatrix} -2 & -2 \\ 4 & 3 \end{pmatrix}$ or $\begin{pmatrix} -1 & -1 \\ 2 & 1\cdot5 \end{pmatrix}$, either form being acceptable.

Now find the inverse of each of the following matrices:

a) $\begin{pmatrix} 5 & 3 \\ 8 & 5 \end{pmatrix}$ b) $\begin{pmatrix} 4 & -3 \\ -1 & 1 \end{pmatrix}$ c) $\begin{pmatrix} 4 & 3 \\ 6 & 5 \end{pmatrix}$ d) $\begin{pmatrix} 2 & -1 \\ 4 & 1 \end{pmatrix}$ e) $\begin{pmatrix} -2 & 1 \\ 3 & 2 \end{pmatrix}$ f) $\begin{pmatrix} 1 & 2 \\ 2 & 1 \end{pmatrix}$

7 The process of finding the inverse of a matrix is also known as 'inverting' the matrix. Invert the following matrices:

a) $\begin{pmatrix} 6 & 7 \\ 3 & 4 \end{pmatrix}$ b) $\begin{pmatrix} 5 & 8 \\ 4 & 7 \end{pmatrix}$ c) $\begin{pmatrix} 3 & 4 \\ 7 & 9 \end{pmatrix}$ d) $\begin{pmatrix} 2 & 1 \\ -2 & 1 \end{pmatrix}$ e) $\begin{pmatrix} 4 & 3 \\ 1 & -1 \end{pmatrix}$ f) $\begin{pmatrix} -2 & 1 \\ 1 & 2 \end{pmatrix}$

g) $\begin{pmatrix} 0 & 3 \\ 1 & 2 \end{pmatrix}$ h) $\begin{pmatrix} 3 & 0 \\ 1 & 4 \end{pmatrix}$ i) $\begin{pmatrix} 2 & 1 \\ 1 & 1 \end{pmatrix}$

In each case, if A is the given matrix, check that $A^{-1}A = AA^{-1} = I$.

8 *Matrix Coding*

a) Write the following message in simple numerical code, using 1 for A, 2 for B.....26 for Z, 27 for a space, 28 for a full stop, 29 for ' and 30 for '.
Message The red cow.   The coded message starts 20 8 5 27.....

b) Arrange the coded message as a (2×2) matrix in the order shown.

It starts $\begin{pmatrix} 20 & 5 & \dots \\ \downarrow & \nearrow \downarrow & \nearrow \\ 8 & 27 & \end{pmatrix}$.

Pre-multiply this matrix by the coding matrix $\begin{pmatrix} 3 & 4 \\ 2 & 3 \end{pmatrix}$.

Write down the product matrix. It starts $\begin{pmatrix} 92 & 123 & \dots \\ 64 & 91 & \dots \end{pmatrix}$.

c) Write down the numbers in this matrix in a straight line, starting
92 64 123 91 .... This is the final form of the coded message ready to send
away.

9 Decoding the Message
a) Here is the coded message:
54 39 89 62 85 57 58 40 93 63 60 43 124 89

Write it as a 'two-row' matrix beginning $\begin{pmatrix} 54 & 89 & \dots \\ 39 & 62 & \dots \end{pmatrix}$.

Pre-multiply it by the 'decoding' matrix $\begin{pmatrix} 3 & -4 \\ -2 & 3 \end{pmatrix}$.

Write down the product matrix beginning $\begin{pmatrix} 6 & 19 & \dots \\ 9 & 8 & \dots \end{pmatrix}$.

b) Write this as a row beginning 6 9 19 8.....
Decode this using the simple numerical code from question 8.

10 a) What is the relation between the coding matrix in question 8 and the decoding
matrix in question 9?
Make up a coding matrix for yourself. It is better if it contains no negatives and if
its determinant is 1, but neither of these conditions is essential. Write down the
decoding matrix.
b) What is the product of the two matrices?
c) You are given three coding matrices. State the decoding matrices.

$\begin{pmatrix} 3 & 1 \\ 2 & 1 \end{pmatrix}$ $\begin{pmatrix} 5 & 2 \\ 2 & 1 \end{pmatrix}$ $\begin{pmatrix} 4 & 1 \\ 7 & 2 \end{pmatrix}$

11 Put the following sentences into code using the matrices stated.

a) The cat drinks the milk. $\begin{pmatrix} 3 & 4 \\ 2 & 3 \end{pmatrix}$

b) 'A stitch in time saves nine.' $\begin{pmatrix} 5 & 7 \\ 2 & 3 \end{pmatrix}$

c) Every cloud has a silver lining. $\begin{pmatrix} 2 & 7 \\ 1 & 4 \end{pmatrix}$

12 Decode the following using the decoding matrix $\begin{pmatrix} 2 & -1 \\ -5 & 3 \end{pmatrix}$.

a) 66 113 81 144 27 51 46 83 75 130 56 96 33 61 104 181
59 100 69 124

b) 21 38 26 48 110 190 93 159 38 67 42 79 44 75 70 126

c) 29 50 100 173 12 23 39 74 96 163 45 84 17 33 90 155 47
80 84 141 15 29 55 98

d) 65 110 48 84 101 175 29 50 66 119 57 105 66 119 17 30
66 113 35 63 41 73 48 89 77 135 97 167 24 47 55 98 114
200

(The four lines together form a well-known clerihew.)

13 Choose a coding matrix, write a message in code to your neighbour, give him or her the decoding matrix and ask for a reply in code.

✻*14* Try and find the inverses of the following matrices.

a) $\begin{pmatrix} 4 & 2 \\ 2 & 1 \end{pmatrix}$ b) $\begin{pmatrix} 3 & -1 \\ 6 & -2 \end{pmatrix}$ c) $\begin{pmatrix} 2 & 3 \\ 4 & 6 \end{pmatrix}$ d) $\begin{pmatrix} 0 & 5 \\ 0 & 4 \end{pmatrix}$

In each case there is no inverse. Why not?

✻*15* When the determinant of a matrix is zero, the matrix is said to be *singular*. All four matrices in question 14 are singular. Comparing the first and second lines of the matrices in a), b) and c) what do you notice? Is the same true for d)?
State which of the following matrices are singular.

e) $\begin{pmatrix} 4 & 3 \\ 8 & -6 \end{pmatrix}$ f) $\begin{pmatrix} 2 & 1 \\ -2 & -1 \end{pmatrix}$ g) $\begin{pmatrix} 3 & 5 \\ 6 & 10 \end{pmatrix}$ h) $\begin{pmatrix} 3\cdot5 & 3 \\ 2\cdot5 & 2 \end{pmatrix}$ i) $\begin{pmatrix} -1 & 0 \\ 2 & 0 \end{pmatrix}$

16 Using Δ for the determinant of the matrix $\begin{pmatrix} p & q \\ r & s \end{pmatrix}$ write down the inverse of this matrix in two different ways.

17 Write down the inverse of each of the following matrices.

a) $\begin{pmatrix} 5 & 3 \\ 8 & 5 \end{pmatrix}$ b) $\begin{pmatrix} 3 & 10 \\ 2 & 7 \end{pmatrix}$ c) $\begin{pmatrix} 3 & 3 \\ 4 & 5 \end{pmatrix}$ d) $\begin{pmatrix} 1 & 0\cdot5 \\ -1 & 0\cdot5 \end{pmatrix}$ e) $\begin{pmatrix} -2 & 1 \\ 1 & 1 \end{pmatrix}$

Check in each case that $MM^{-1}=M^{-1}M=I$.

✻*18* Where possible, find the inverse of each of the following matrices. Where it is not possible, give a reason.

a) $\begin{pmatrix} 2 & 5 \\ 3 & 8 \end{pmatrix}$ b) $\begin{pmatrix} 3 & 1 \\ 1 & -1 \end{pmatrix}$ c) $\begin{pmatrix} 0\cdot2 & 1\cdot2 \\ 0\cdot7 & -0\cdot8 \end{pmatrix}$ d) $\begin{pmatrix} 3 & -2 \\ 8 & -6 \end{pmatrix}$ e) $\begin{pmatrix} 2 & 3 \\ 7 & 8 \end{pmatrix}$

f) $\begin{pmatrix} -1 & -4 \\ 3 & 12 \end{pmatrix}$ g) $\begin{pmatrix} 0\cdot75 & -0\cdot25 \\ 8 & 4 \end{pmatrix}$ h) $\begin{pmatrix} 0 & 3 \\ 1 & 0 \end{pmatrix}$ i) $\begin{pmatrix} 2 & 0 \\ -3 & 0 \end{pmatrix}$ j) $\begin{pmatrix} 3\cdot6 & 0\cdot2 \\ 2\cdot0 & 0\cdot25 \end{pmatrix}$

6C   The Solution of Simultaneous Equations by Matrix Inversion

Either work numbers 1 to 4 and then 7 onwards *or* start at number 5.

1   The simultaneous equations $\left.\begin{array}{l} 3x+2y=9 \\ x+\ y=4 \end{array}\right\}$ can be written in matrix form as:

$\begin{pmatrix} 3 & 2 \\ 1 & 1 \end{pmatrix}\begin{pmatrix} x \\ y \end{pmatrix}=\begin{pmatrix} 9 \\ 4 \end{pmatrix}$

This is because the matrix product on the left is $\begin{pmatrix} 3x+2y \\ x+\ y \end{pmatrix}$ and as this has to equal the matrix on the right, $3x+2y$ must equal 9 and $x+y$ must equal 4.

a) Write down the inverse of the matrix $\begin{pmatrix} 3 & 2 \\ 1 & 1 \end{pmatrix}$.

b) Pre-multiply both sides of the equation by this inverse.
The product of a matrix and its inverse is I, so the left side of the equation is $\begin{pmatrix} x \\ y \end{pmatrix}$. What is the right side?

c) Equating individual terms in the two matrices on the right and left gives the separate values of x and y. What are they?

2 Use the method of question 1 to solve $\begin{array}{l}3x+2y=4\\7x+5y=9\end{array}\Big\}$.

3 Use the same method to solve $\begin{array}{l}4x-3y=1\\x-2y=4\end{array}\Big\}$.

4 Using the inverse matrix method, solve the following pairs of equations:

a) $\begin{array}{l}5x-2y=0\\8x-3y=-1\end{array}\Big\}$

b) $\begin{array}{l}3x-2y=7\\2x-y=4\end{array}\Big\}$

c) $\begin{array}{l}6x-4y=6\\2x-y=1\end{array}\Big\}$

d) $\begin{array}{l}5x+6y=-4\\3x+4y=-2\end{array}\Big\}$

e) $\begin{array}{l}5x-3y=12\\2y-3x=-7\end{array}\Big\}$

5 *An Alternative Approach*

The equations $\begin{array}{l}x-3y=7\\2x+3y=-4\end{array}\Big\}$ can be written in matrix form as $\begin{pmatrix}1&-3\\2&3\end{pmatrix}\begin{pmatrix}x\\y\end{pmatrix}=\begin{pmatrix}7\\-4\end{pmatrix}$

or $AX=B$ where A is the square matrix of coefficients $\begin{pmatrix}1&-3\\2&3\end{pmatrix}$, X is the column

matrix of variables $\begin{pmatrix}x\\y\end{pmatrix}$, and B is the column matrix of constants $\begin{pmatrix}7\\-4\end{pmatrix}$.

Write in full the values of A, X and B for each of the following pairs of equations:

a) $\begin{array}{l}2x-9y=-5\\3x+2y=8\end{array}$

b) $\begin{array}{l}7y+8z=11\\3y-2z=2\end{array}$

c) $\begin{array}{l}8c-5w=3\\2c+3w=5\end{array}$

d) $\begin{array}{l}5x+3y=11\\2x-y=0\end{array}$

e) $\begin{array}{l}a-7b-9=0\\2a+b-3=0\end{array}$

f) $\begin{array}{l}q+3x-7=0\\3q-6x+9=0\end{array}$

6 Pre-multiplying the matrix equation $AX=B$ by A^{-1} (the inverse of A) gives $A^{-1}AX=A^{-1}B$, i.e. $IX=A^{-1}B$, i.e. $X=A^{-1}B$.

Thus for the equations $\begin{array}{l}x-2y=1\\2x+y=7\end{array}\Big\}$ A is $\begin{pmatrix}1&-2\\2&1\end{pmatrix}$ and A^{-1} is $\frac{1}{5}\begin{pmatrix}1&2\\-2&1\end{pmatrix}$

so $\begin{pmatrix}x\\y\end{pmatrix}=\frac{1}{5}\begin{pmatrix}1&2\\-2&1\end{pmatrix}\begin{pmatrix}1\\7\end{pmatrix}=\frac{1}{5}\begin{pmatrix}15\\5\end{pmatrix}=\begin{pmatrix}3\\1\end{pmatrix}$ i.e. $x=3$, $y=1$.

Now solve the following equations by the above method, the method of 'matrix inversion':

a) $\begin{array}{l}2x+y=10\\3x+2y=17\end{array}$

b) $\begin{array}{l}3x+2y=0\\4x+3y=-1\end{array}$

c) $\begin{array}{l}2x+y=11\\x+4y=23\end{array}$

d) $\begin{array}{l}3x-2y=-1\\7x-5y=-3\end{array}$

e) $\begin{array}{l}5x+4y=21\\3x+2y=11\end{array}$

f) $\begin{array}{l}4x+2y=2\\8x+5y=3\end{array}$

7 Rearrange the following equations and solve them by matrix inversion:

a) $\begin{array}{l}2x=9-y\\2y=14-3x\end{array}$

b) $\begin{array}{l}5=3x+2y\\y=2-x\end{array}$

c) $\begin{array}{l}2y=4(1-x)\\3(y-2)+2x=0\end{array}$

d) $\begin{array}{l}2x=4-3y\\2=x+5y\end{array}$

e) $\begin{array}{l}2x=-3y+1\\y=5-3x\end{array}$

f) $\begin{array}{l}4+2y=2x\\y=10-3x\end{array}$

8 Solve the following equations by matrix inversion:

a) $\begin{array}{l}7a+b=9\\3a+2b=7\end{array}$

b) $\begin{array}{l}11p+6q=62\\5p+3q=29\end{array}$

c) $\begin{array}{l}4g-7h=-12\\2g+5h=11\end{array}$

d) $\begin{array}{l}2x=3y+8\\5x+7y+9=0\end{array}$ *e)* $\begin{array}{l}3z-w-4=0\\4z=22-7w\end{array}$ *f)* $\begin{array}{l}2m=n-1\\5m+28=11n\end{array}$

✳9 If the coefficient matrix is singular (i.e. if its determinant is zero) there is no inverse matrix and the equations cannot be solved by matrix inversion.

Which of the following pairs of equations cannot be solved by matrix inversion?

a) $\begin{array}{l}3x+y=1\\6x+2y=7\end{array}$ *b)* $\begin{array}{l}x-4y=2\\0{\cdot}5x-2y=1\end{array}$ *c)* $\begin{array}{l}x+2y=5\\2x+y=4\end{array}$

d) $\begin{array}{l}3x+y=7\\6x-y=11\end{array}$ *e)* $\begin{array}{l}5x-2y=7\\-x+0{\cdot}4y=-1{\cdot}4\end{array}$

✳✳10 When the coefficient matrix is singular there is either no solution or an infinity of solutions.

In 9a) there is no solution. Any values of x and y that satisfy the first equation cannot possibly satisfy the second. The second equation is 'incompatible' with the first.

In 9e) there is an infinity of solutions, such as 3,4 or 5,9 or 7,14, etc. The second equation is 'redundant' (i.e. an exact multiple of the first).

In the following pairs of equations, if there is a unique solution give it. If there is no solution, say so. If there is an infinity of solutions, give one or two specimen solutions.

a) $\begin{array}{l}3x+4y=7\\2x-y=1\end{array}$ *b)* $\begin{array}{l}3x+4y=7\\6x+8y=14\end{array}$ *c)* $\begin{array}{l}3x+4y=7\\6x+8y=11\end{array}$

d) $\begin{array}{l}2x-y=3\\6x-3y=0\end{array}$ *e)* $\begin{array}{l}3x+y=7\\x-y=1\end{array}$ *f)* $\begin{array}{l}x-2y=6\\3x-6y=18\end{array}$

✳ 6D More about Determinants

An array of numbers such as $\begin{array}{cc}1&3\\2&2\end{array}$ can be either a determinant, written $\begin{vmatrix}1&3\\2&2\end{vmatrix}$, or a matrix, written $\begin{pmatrix}1&3\\2&2\end{pmatrix}$.

A determinant can be reduced to a single number, which in the case above is -4. A matrix cannot be reduced to a single number. It is an operator, meaning 'do something'. (The 'determinant of a matrix', of course, is a single number, as we have already seen.)

1 The value of the 2×2 determinant $\begin{vmatrix}a&b\\c&d\end{vmatrix}$ is $ad-bc$.

(The value of a 3×3 determinant such as $\begin{vmatrix}1&2&3\\2&4&5\\1&3&6\end{vmatrix}$ is much harder to calculate. An example is given in Chapter 23B.)

Calculate the values of the following 2×2 determinants:

a) $\begin{vmatrix}3&1\\2&6\end{vmatrix}$ *b)* $\begin{vmatrix}1&0\\-1&1\end{vmatrix}$ *c)* $\begin{vmatrix}0&-3\\-2&6\end{vmatrix}$ *d)* $\begin{vmatrix}\frac{1}{4}&\frac{3}{4}\\\frac{1}{3}&\frac{2}{3}\end{vmatrix}$ *e)* $\begin{vmatrix}0{\cdot}2&0{\cdot}6\\1{\cdot}0&3{\cdot}0\end{vmatrix}$

f) $\begin{vmatrix}p&q\\r&s\end{vmatrix}$ *g)* $\begin{vmatrix}x&-2\\2x&3x\end{vmatrix}$ *h)* $\begin{vmatrix}(x-1)&(x-2)\\(x+3)&(x-5)\end{vmatrix}$ *i)* $\begin{vmatrix}a^2&bc\\bc&ab\end{vmatrix}$ *j)* $\begin{vmatrix}2\pi&\pi r\\-2h&r^2\end{vmatrix}$

2 Determinants have many uses, one such being the calculation of areas from a given set of co-ordinates, an example of which is given in Chapter 23B. They also give an extremely quick method for the solution of simultaneous equations in two variables.

This method is known as Crout's method.

Step 1   Write the equations with all the terms on one side, e.g.

$$x+2y-4=0$$
$$2x+3y-7=0$$

Step 2 $\dfrac{x}{\begin{vmatrix}2 & -4\\3 & -7\end{vmatrix}}=\dfrac{-y}{\begin{vmatrix}1 & -4\\2 & -7\end{vmatrix}}=\dfrac{1}{\begin{vmatrix}1 & 2\\2 & 3\end{vmatrix}}$

The determinant under x is obtained from the two columns not containing x.
The determinant under y is obtained from the two columns not containing y.
The determinant under 1 is obtained from the two columns containing x and y.

Step 3   Write in the values of the determinants.

$$\frac{x}{-2}=\frac{-y}{1}=\frac{1}{-1}$$

What is the final solution?

3   Use Crout's method to solve the following equations:

a) $\begin{aligned}2x+3y-10&=0\\x-2y+2&=0\end{aligned}$ b) $\begin{aligned}x-3y+5&=0\\2x+y-11&=0\end{aligned}$

c) $\begin{aligned}2x-y-3&=0\\x+2y-4&=0\end{aligned}$ d) $\begin{aligned}3x-2y&=3\\2x+y&=-5\end{aligned}$

4   One advantage of Crout's method is that with a little practice you can leave out Step 2 altogether and go straight to Step 3, working the value of the determinants in your head.
Solve the following equations, writing down only Step 3 and the final answer:

a) $\begin{aligned}x-y-1&=0\\x-2y+2&=0\end{aligned}$ b) $\begin{aligned}x-3y+5&=0\\2x+y-11&=0\end{aligned}$

c) $\begin{aligned}2x-y-3&=0\\x+2y-4&=0\end{aligned}$ d) $\begin{aligned}4x+y-3&=0\\x-2y&=3\end{aligned}$

✱✱5   In the following equations, one of each pair is either redundant or incompatible (see 6C question 10). What happens when you try to apply Crout's method?

a) $\begin{aligned}3x+2y&=6\\6x+4y&=12\end{aligned}$ b) $\begin{aligned}3x+2y&=6\\6x+4y&=10\end{aligned}$ c) $\begin{aligned}x-3y&=5\\0{\cdot}2x-0{\cdot}6y&=1\end{aligned}$

7 More Quadratic Equations

7A Completing the Square

1 Solve the equations:

a) $(x+1)^2 = 4$ b) $(x+3)^2 = 16$ c) $(x-2)^2 = 3$

d) $(x-3)^2 = 7$ e) $(2x-3)^2 = 25$ f) $(2x+1)^2 = 10$

Remember there are two answers for each equation.

2 Write each of the following in full:

a) $(x+1)^2$ b) $(x+2)^2$ c) $(x+5)^2$ d) $(x-4)^2$ e) $(x-7)^2$

3 Complete the following:

a) $x^2+8x \quad =(x+\ )^2$ b) $x^2-8x \quad =(x-\ )^2$

c) $x^2+6x \quad =(x\ )^2$ d) $x^2-6x \quad =(x\ )^2$

e) $x^2-2x \quad =(\ )^2$ f) $x^2+x \quad =(\ \frac{1}{2})^2$

g) $x^2+5x \quad =(x+\ )^2$ h) $x^2-3x \quad =(x\ )^2$

i) $2x^2-3x \quad =2(\ )^2$ j) $3x^2-8x \quad =3(\ )^2$

4 Solve the equation $x^2+2x=5$ by adding the same number to both sides and choosing this number so as to make the left-hand side a perfect square.

5 Solve the equation $x^2-4x=1$ by the method of question 4.

6 Solve each of the following equations by the method of question 4, which is known as 'completing the square'. Give your answers correct to 2 decimal places.

a) $x^2-2x=1$ b) $x^2+6x=3$ c) $x^2+x=4$

d) $x^2+4x+2=0$ e) $x^2+3x+1=0$ f) $x^2-5x+5=0$

g) $x^2-3x-5=0$ h) $2x^2+4x-3=0$ i) $2x^2+5x+3=0$

What must you do to h) and i) before attempting to complete the square?

✱7 Using the method of completing the square, find two expressions for x which satisfy the equation $ax^2+bx+c=0$.

(*Hint* Take c to the other side, divide through by a and then complete the square.)

7B Solution of Quadratic Equations by Formula

In this section give your answers (where appropriate) correct to 2 decimal places.

1 Question 7 above gave you the formula for solving the quadratic equation

$ax^2+bx+c=0$ as $x=\dfrac{-b\pm\sqrt{b^2-4ac}}{2a}$.

Using this formula, solve the following equations:

a) $x^2+4x-6=0$ b) $x^2-3x-7=0$ c) $x^2+6x+8=0$

d) $x^2-x-3=0$ e) $x^2-5x+3=0$ f) $x^2+9x+5=0$

2 The two roots in question 1 are $\dfrac{-b+\sqrt{b^2-4ac}}{2a}$ and $\dfrac{-b-\sqrt{b^2-4ac}}{2a}$.

These add to give $-2b/2a$ or $-b/a$. It will be shown in Chapter 17 that they multiply to give c/a. This gives a useful check on the roots of quadratic equations. The sum of the roots is $-b/a$ and the product of the roots is c/a.

The first part of the check (i.e. the sum of the roots) is very simple, and should always be applied.

Apply the full check to your answers in question *1 a, b* and *c*.

3 Solve each of the following:

a) $x^2 + 7x + 4 = 0$ b) $2x^2 - 5x - 4 = 0$ c) $2x^2 + 9x + 3 = 0$
d) $3x^2 + 2x - 3 = 0$ e) $5x^2 + x - 2 = 0$ f) $5x^2 + 8x + 2 = 0$

Check *a* and *b* using the check given in question 2.

4 Solve:

a) $x^2 + 10x = 5$ b) $x^2 - 8x + 6 = 0$ c) $2x^2 - x = 7$
d) $4x^2 + 3x = 5$ e) $5x^2 = 7x - 1$ f) $3x^2 - 9x + 5 = 0$.

Check *e* and *f*.

5 Find the solutions of the following equations:

a) $x^2 + 2x - 5 = 0$ b) $x^2 - 15x + 8 = 0$ c) $2x^2 - x - 4 = 0$
d) $5x^2 + 9x + 2 = 0$ e) $3x^2 = 7x - 3$ f) $2x^2 + 11x + 8 = 0$

6 Solve:

a) $x(x - 2) = 5$ b) $2x(x + 3) + 3 = 0$ c) $(x - 2)(x - 4) = 5$
d) $(x + 1)(x - 5) = 3$ e) $(2x + 1)(x - 1) = 1$ f) $(2x - 5)(x + 3) = 8$

7 Solve:

a) $x^2 - 8x = 7$ b) $x^2 + 5x - 2 = 0$ c) $3x^2 + 8x + 3 = 0$
d) $2x^2 = 7x + 3$ e) $5x^2 + 9x - 3 = 0$ f) $4x^2 - 9x = 3$
g) $x^2 - 10x + 8 = 0$ h) $x^2 + 12x + 10 = 0$ i) $x^2 = 8x - 11$
j) $2x^2 = 5 - 7x$ k) $3x^2 + 8 = 14x$ l) $5x^2 - 2 = 6x$

8 Solve:

a) $(x - 2)(x - 5) = 5$ b) $x + 4 = x(x - 3)$

c) $\dfrac{1}{x + 5} = x + 3$ (*Hint*   Multiply both sides of the equation by $x + 5$.)

d) $x = \dfrac{5}{x + 2}$ (*Hint*   Multiply both sides of the equation by $x + 2$.)

e) $\dfrac{7x}{x - 2} = x + 4$ f) $\dfrac{x(x + 7)}{x - 2} = 2$

9 Solve:

a) $\dfrac{2x}{x + 1} - \dfrac{1}{x - 2} = 0$ (*Hint*   Multiply both sides of the equation by $(x + 1)(x - 2)$.)

b) $\dfrac{2}{x} + \dfrac{1}{x - 5} = 3$ c) $\dfrac{4x}{x + 3} - \dfrac{x}{x - 4} = 2$

d) $\dfrac{5}{x + 2} = 4 + \dfrac{1}{x - 8}$ e) $\dfrac{4x}{(x - 1)^2} - \dfrac{3}{x - 1} = 2$

10 Solve:

a) $\dfrac{3}{x+1}=1+\dfrac{2}{x+5}$ b) $\dfrac{1}{x+1}-\dfrac{2}{x+2}=3$

c) $\dfrac{2}{x^2-4}=\dfrac{x}{x+2}$ (*Hint* Factorise x^2-4 first.)

d) $\dfrac{3x}{x^2-25}=1-\dfrac{3}{x+5}$ e) $\dfrac{4}{x^2-2x-3}=\dfrac{x}{x-3}$

11 Solve:

a) $\dfrac{4}{x-6}+2=\dfrac{x}{x+4}$ b) $\dfrac{1}{x}+\dfrac{2}{x-3}=5$ c) $\dfrac{3x}{x+3}=\dfrac{1}{x^2-9}$

d) $\dfrac{3}{x^2-5x+6}=\dfrac{4}{x-3}-1$ e) $\dfrac{1}{x+4}-\dfrac{2}{x-1}=\dfrac{3}{x+2}$

7C Miscellaneous Quadratic Equations

Solve each of the following equations. You should be able to solve some of the quadratic equations by finding factors. If you have to use the formula, give your answers correct to 2 d.p.

1 $x^2-5x=36$

2 $(x-3)(x+2)=6$

3 $\dfrac{1}{x-2}-\dfrac{3}{x+1}=1$

4 $\dfrac{2x}{x+3}=1+\dfrac{1}{x-2}$

5 $2x^2-7x-6=0$

6 $(x-2)(x-5)=10$

7 $\dfrac{3x}{x-2}+\dfrac{1}{x^2-4}=0$

8 $\dfrac{1}{x+1}+\dfrac{1}{x-3}=\dfrac{1}{x-4}$

9 $x^2-12x+7=0$

10 $x^2+9x+9=0$

11 $4x^2-36=0$

12 $3x^2-75x=0$

13 $5x^2=1-7x$

14 $5x^2=7x-2$

15 $5x(x-2)=15$

16 $4=3x+2x^2$

17 $6x^2-7x-5=0$

18 $12x^2-x=6$

19 $10x^2-3x-2=0$

20 $5x^2+9x-2=0$

21 Find the length of the hypotenuse of a right-angled triangle if it is 2 cm longer than one side and 4 cm longer than the other side.

22 a) 43 is written as 133 in base *m*. Find *m*.
  b) 889 is written as 379 in base *n*. Find *n*.

23 The sum of the squares of three consecutive numbers is 149. Find the numbers. (*Hint* Take *x* as the middle number.)

24 The product of two consecutive numbers is 12 more than 5 times the smaller. Find the two numbers.

25 Travelling at an average speed of x km/h on a journey of 180 km and at an average speed of 15 km/h less on the return journey, the total time taken is 7 hours. Find the average speed on the outward journey.

26 22 bars of chocolate can be bought for £1.44. 84p of this is spent on filled bars and the rest of the money is spent on plain bars costing 1p less per bar than the filled bars. Find the price of one of the plain bars.

27 The volume of a prism on a square base is increased by 50% when the length of the sides of the base is increased by 3 cm. Find the length of the sides of the base to the nearest tenth of a cm.

28 Two rectangles have the same width. The length of the larger one is twice its width and the length of the other is 2 cm more than its width. The difference in their areas is 5 cm^2. Find their width.

29 The length of a rectangle is 6 cm more than its width. If the length of the rectangle is increased by 40% and the width is decreased by 20%, the area is increased by 3 cm^2. Find the length of the original rectangle.

30 If a sum of money is shared equally among two more people than was originally intended, the share received by each of the original people is decreased by one fortieth of the total sum. Find the original number of people.

7D Simultaneous Quadratic Equations

1 If $x^2 - y^2 = 20$ and $x + y = 10$, find the value of $x - y$. You now have two simultaneous linear equations in x and y. Solve them and find the values of x and y.

2 If $4x^2 - y^2 = 96$ and $2x - y = 8$, find the value of $2x + y$. Use this to find the values of x and y.

3 Solve the following pairs of equations:

a) $\begin{aligned} x^2 - 9y^2 &= 45 \\ x - 3y &= 15 \end{aligned}$
   b) $\begin{aligned} x^2 - y^2 &= 39 \\ x + y &= -13 \end{aligned}$
   c) $\begin{aligned} 4x^2 - 25y^2 &= 119 \\ 2x + 5y &= 7 \end{aligned}$

d) $\begin{aligned} 3x - y &= 11 \\ 3x^2 - xy &= 22 \end{aligned}$
   e) $\begin{aligned} x - 2y &= 11 \\ 2y^2 &= 44 + xy \end{aligned}$
   f) $\begin{aligned} 2x - y &= 5 \\ 4x^2 - 2xy &= 70 \end{aligned}$

4 Solve the following pair of equations by substituting for y in terms of x:
$$x^2 + 2xy + 8 = 0$$
$$y = 1 - 2x$$

5 Find values of x and y which satisfy these equations: $\begin{aligned} x^2 - 2y^2 - 3x &= 2 \\ y &= x - 3 \end{aligned}$.

6 Solve:

a) $\begin{aligned} y^2 &= 3x^2 - 4xy + 2 \\ x + y &= 4 \end{aligned}$
   b) $\begin{aligned} x^2 - y^2 &= 21 \\ x &= 2y + 1 \end{aligned}$
   c) $\begin{aligned} x + 2y &= 3 \\ x^2 - 2xy + x &= 24 \end{aligned}$
   (Substitute for y.)

d) $\begin{array}{l} x-2y+7=0 \\ 3x-xy+y^2=19 \end{array}$ e) $\begin{array}{l} xy=12 \\ x^2+y^2=25 \end{array}$ f) $\begin{array}{l} 2xy=5 \\ x+4y+3=14 \end{array}$
(Substitute for x.)

7 The hypotenuse of a right-angled triangle is 17 and the sum of the lengths of the sides is 23. What are the lengths of the sides?

8 The external perimeter of the figure in the diagram is 38 and the overall area 90. Find x and y.

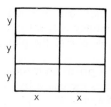

9 Repeat question 8 for a perimeter of 200 and an area of 650. Give your answers to 3 s.f.

10 A boy has a piece of thin wire 96 cm long. He cuts it into two pieces and bends each piece to form a square. If the sum of the areas of the two squares is 338 cm², where did he cut the wire?

11 A grandmother gives her grandchildren a sum of money each birthday. On their tenth birthday she gives them 10p for every year of their life. On their eleventh birthday she gives them 11p for every year of their life.

a) How much does she give a 16-year-old on his or her 16th birthday?
b) If the combined ages of two grandchildren are 28 years and she gives them £3.94 altogether, how old is each?

✳12 A stone is catapulted vertically upwards from the edge of a cliff. After x seconds its height above the edge of the cliff is $20x-5x^2$. Two seconds later another stone is dropped from the edge of the cliff, and after y seconds its depth below the edge of the cliff is $5y^2$. Find the value of x when the stones are 20 m apart.

7E Graphical Solution of Equations

1 Draw the graph of $y=x^2-5x+6$ for $-2\leqslant x\leqslant 6$. Use your graph to solve   a) $x^2-5x+6=0$ b) $x^2-5x=0$ (i.e. $x^2-5x+6=6$)
c) $x^2-5x-4=0$ d) $x^2-5x+2=0$.
e) On the same axes draw the line $y=x$ and hence find the solutions of the equation $x^2-6x+6=0$.

2 Draw the graph of $y=x^2+3x-1$ for $-5\leqslant x\leqslant 2$. On the same axes draw the line $y=x+3$, and hence solve the equation $x^2+2x-4=0$. Check your answers by calculation.

3 Taking values of x from -4 to $+4$, draw on the same axes $y=6/x$ and $y=x+1$. Show how your graph helps you to solve the equation $x^2+x=6$, and write down the solution.

4 Draw the graph of $y=x^3-x^2-2x$ for $-3\leqslant x\leqslant 5$. Use your graph to solve the equations   a) $x^3-x^2-2x=0$ b) $x^3-x^2=2x-2$.
State the range of values for which $x^3-x^2-2x<0$.

5 Taking values of x between -2 and $+6$, draw the graph of $y=x^3-5x^2+3x+1$. Use your graph to solve the equations   a) $x^3-5x^2+3x+1=0$
b) $x^3-5x^2+3x+6=0$.
By drawing the line $y=k$, you can solve the equation $x^3-5x^2+3x+1-k=0$. For some values of k this equation only has one real root. Give the range of values of k for which this is true. Give the range of values of k for which the equation has three positive real roots.

6 Draw the graph of $y=x^2-2x-3$ for $-2\leqslant x\leqslant 6$. On the same axes draw the line $y=2x-5$ and hence solve the equation $x^2-4x+2=0$. Explain the results when you draw the line $y=2x-7$. What equation are you solving? What do you know about the roots of the equation $x^2-4x+5=0$.

7 Draw the graph of $y=(1+x)(7-2x)$ for $-2\leqslant x\leqslant 6$. By drawing a suitable line, find for what range of values $(1+x)(7-2x)>3x$. Find the range of values for k such that the equation $(1+x)(7-2x)=k$ has one negative root and one positive root between 2 and 3.

8 Draw the graph of $y=1/x+x$ for $0\cdot5\leqslant x\leqslant 5$.
Use your graph to find solutions to the equation $x^2-3x+1=0$.
Find the range of values for k for which the equation $\dfrac{1}{x}+x=k$ has no real roots for values of x between $0\cdot5$ and 5.

∗9 The diagram shows the graph of $x^2+y^2=25$. If P is a point on the curve, and the co-ordinates of P are (x, y), then $OP^2=x^2+y^2$, i.e. $OP^2=25$. So the curve is a circle, centre O, radius 5.
Draw the graph and use it to solve the equations:

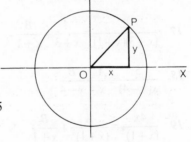

a) $\begin{array}{l} x^2+y^2=25 \\ x+y=7 \end{array}$ b) $\begin{array}{l} x^2+y^2=25 \\ x-y=1 \end{array}$ c) $\begin{array}{l} x^2+y^2=25 \\ x=2y \end{array}$

∗10 Draw the curve $y=24/x$ from $x=2$ to $x=10$. Use your curve to solve the equations:

a) $\begin{array}{l} xy=24 \\ x+y=10 \end{array}$ b) $\begin{array}{l} xy=24 \\ y-x=5 \end{array}$ c) $\begin{array}{l} xy=24 \\ 3x+2y=30 \end{array}$

∗7F   Identities

Note on identity and equation   An equation in x is true only for certain values of x. An identity in x is true for all values of x. Thus $x^2-x=6$ is an equation. It is true for $x=3$ or -2 and for no other values of x. But $x^2-2x+1\equiv(x-1)^2$ is an identity. It is true for any value of x.

In each of the following find the values of the constants $A, B, C\ldots$.

Example   In No. 1 the LHS is $x^2-2x-35$, so $A=1, B=-2$, etc.

1 $(x+5)(x-7)\equiv Ax^2+Bx+C$ **2** $x(x-1)+3x^2\equiv Ax^2+Bx+C$

3 $x(x+2)+x(x-3)\equiv Ax^2+Bx+C$ **4** $(x+4)(x-6)+5x\equiv Ax^2+Bx+C$

5 $4x^2-3-(x+2)(x-5)\equiv Ax^2+Bx+C$ **6** $x^2+8x-5\equiv Ax(x+1)+B(x-1)+C$

63

7 $3x^2 - 7x + 9 \equiv Ax(x+1) + B(x+2) + C$

8 $x^3 + 4x^2 - 5x + 2 \equiv Ax(x-1)(x+1) + Bx(x-1) + Cx + D$

9 $x^3 - 1 \equiv Ax(x+1)(x+2) + B(x+1)(x+2) + C(x+2) + D$

10 $(x+2)(x^2 - 3x - 2) \equiv x^3 + Ax^2 + Bx + C$

11 $(x^2 - 5)(x-2) \equiv x^3 + Ax^2 + Bx + C$

12 $2x^3 - 3x^2 - 5 \equiv Ax(x^2+1) + Bx(x+1) + Cx + D$

13 $3(x-2)^2 - (x-2) \equiv A(x^2+1) + Bx + C$

14 $5x \equiv (Ax+B)(x+1) + C(x^2-2)$

15 $\dfrac{1}{(x-1)(x+2)} \equiv \dfrac{A}{x-1} + \dfrac{B}{x+2}$

(*Hint* Multiply both sides by $(x-1)(x+2)$, so $1 \equiv A(x+2) + B(x-1)$, then continue as before.)

16 $\dfrac{3x}{(x+5)(x-7)} \equiv \dfrac{A}{x+5} + \dfrac{B}{x-7}$

17 $\dfrac{2}{(x+3)(x+1)} \equiv \dfrac{A}{x+3} + \dfrac{B}{x+1}$

18 $\dfrac{5}{x(x-4)} \equiv \dfrac{A}{x} + \dfrac{B}{x-4}$

19 $\dfrac{4x}{(x+1)^2} \equiv \dfrac{A}{(x+1)^2} + \dfrac{B}{x+1}$

20 $\dfrac{2x}{x^2-9} \equiv \dfrac{A}{x+3} + \dfrac{B}{x-3}$

21 Given that $(x-3)$ and (x^2+ax+b) are factors of $x^3 - 8x^2 + 17x - 6$, find the values of a and b.

22 Given that the factors of $x^3 + 3x^2 - 6x - 8$ are $x+4$ and $x^2 + ax + b$, find the values of a and b. Hence find the solutions of the equation $x^3 + 3x^2 - 6x - 8 = 0$.

23 If $x+3$ and $x^2 + ax + b$ are the factors of $x^3 - x^2 - 8x + 12$, find the values of a and b. Hence completely factorise the expression.

24 If $(ax+b)^3 \equiv 8x^3 + cx^2 + 6x + 1$, find the values of a, b and c.

25 If $y+a$ and $y^2 + by + 9$ are the factors of $y^3 - 27$, find the values of a and b.

8 The Area under a Curve

8A

Counting Squares (Revision questions)

1 On graph paper, taking 2 cm as the unit on each axis, draw the graph of $y = 10x - 16 - x^2$ for values of x from 2 to 8. Find the area between the curve and the x axis by counting squares.
It will help you to draw a series of rectangles as shown in the diagram. The number of 'two millimetre squares' in each of these can be found by simple multiplication. The number of two millimetre squares in the remaining area can then be counted. Addition gives the total number in the whole area.

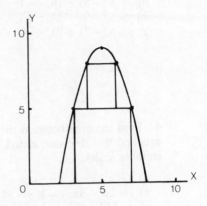

The unit of area must next be determined. It depends on the scale you are using. In this case it is a 'two centimetre square'. Now answer the following questions:

 a) How many two millimetre squares are there in one unit of area?
 b) How many two millimetre squares are there in the total area under the curve?
 c) What is the area under the curve?

2 Repeat question 1 with the same curve but this time 'box it in' and subtract the area between the curve and the box from the area of the box. State:

 a) the area between the curve and the box,
 b) the area of the box,
 c) the area between the curve and the x axis.

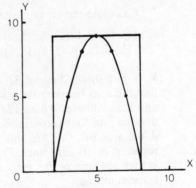

3 *a)* Draw the curve $y = 6x - x^2$ for values of x from 0 to 6. Take 2 cm as the unit on each axis. Find the area between the curve and the x axis.
 b) Draw the curve $y = 10x - x^2 - 9$, between $x = 1$ and $x = 9$, taking 2 cm as the unit on the x axis and 1 cm on the y axis.
 i) Count the number of centimetre squares between the curve and the x axis.
 ii) What is the unit of area? How many centimetre squares are there in one unit of area?
 iii) What is the area between the curve and the x axis?

4 Draw the curve $y = x^2$ between $x = 0$ and $x = 8$, taking 2 cm as the unit on the x axis and 1 cm to 5 units on the y axis.
 a) What is the unit of area?
 b) How many two millimetre squares are there in one unit of area?
 c) What is the area under the curve?

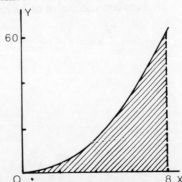

5 Find the area between the curve, the x axis and the ordinates stated. Choose suitable scales.

a) $y = x^2 - 8x + 16$, $x = 1$, $x = 7$
b) $y = x^2 - 9x + 24$, $x = 1$, $x = 8$
c) $y = x^3 - 7x + 10$, $x = 0$, $x = 4$

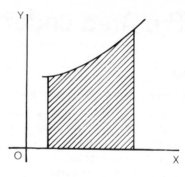

* 6 Find the area between the curve, the y axis and the abscissae stated. Choose suitable scales.

a) $y = x^2$, $y = 4$, $y = 64$ ($x = 2$ to 8)
b) $y = 2x^2 - 3x$, $y = 2$, $y = 54$ ($x = 2$ to 6)
c) $y = 154 - 10x - x^2$, $y = 10$, $y = 130$

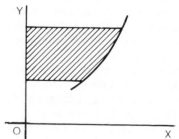

The Trapezium Rule (Revision)

7 Calculate the areas of the following trapeziums whose co-ordinates are given:

a) $(3, 0)$, $(7, 0)$, $(7, 6)$, $(3, 2)$ b) $(1, 2)$, $(1, 8)$, $(4, 7)$, $(4, 2)$
c) $(3, 1)$, $(5, 5)$, $(5, 6)$, $(3, 6)$ d) $(4, 2)$, $(4, 7)$, $(8, 7)$, $(5, 2)$

8 Fig. 1 shows a curve BJ. It is required to find the area between the curve, the x axis and the ordinates AB and IJ. The area is divided into four trapeziums of equal width d, the arcs BD, DF, FH and HJ being replaced by straight lines. The required area is then the sum of the areas of the trapeziums, i.e.

$$\frac{d}{2}[(AB+CD)+(CD+EF)+(EF+GH)+(GH+IJ)]$$

or $d[\frac{1}{2}(AB+IJ)+(CD+EF+GH)]$ or in words, (half the sum of the first and last ordinates plus the sum of the intermediate ordinates) times the constant distance between the ordinates. Apply this rule to find the shaded area in Fig. 2, the distance between successive ordinates being 3.

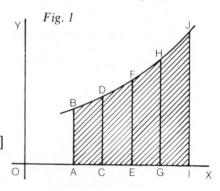

Fig. 1

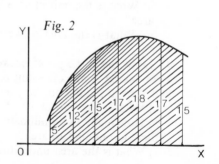

Fig. 2

9 Apply the trapezium rule to find the area between the curve, the x axis and the first and last ordinate in each of the following cases. You are given the ordinates and the constant distance between them.

    a) 3·0, 5·0, 6·8, 8·4, 9·8, 11·0, 12·0, 12·8, 13·6 $d=2$
    b) 12·0, 10·0, 8·3, 6·9, 5·8, 5·0, 4·5, 4·2 $d=3$
    c) 3·5, 6·8, 8·4, 9·8, 11·2, 12·0, 13·0, 13·5 $d=1$

10 Calculate the values of y for the values of x stated. In each case calculate also the area bounded by the curve, the x axis, and the first and last of the given ordinates. (Use the trapezium rule.)

    a) $y=x^2-7x+16$ $x=0, 1,\ldots 7$
    b) $y=7x-x^2+6$ $x=0, 1,\ldots 7$
    c) $y=x^3-10x^2+27x+10$ $x=1, 2,\ldots 6$
    d) $y=x^3-6x^2+8x+6$ $x=0, 1,\ldots 5$

* 11 Given that $y=12x-x^2-4$, calculate the values of y for $x=0, 1, 2\ldots 12$. Using the trapezium rule, calculate the area between the curve, the x axis and the first and last ordinates for the following values of w, the width of the trapeziums:

    a) $w=4$ *b)* $w=3$ *c)* $w=2$ *d)* $w=1$

Given that the true area is 336, what can you deduce from your four answers?

Simpson's Rule

12 Simpson's rule gives a closer approximation than the trapezium rule to the area under a curve. The area is divided into a number of strips as shown. The number of strips must be *even*. All the strips are of equal width h. The ordinates are lettered $y_0, y_1, y_2 \ldots y_n$. The area is then given by $A=\frac{1}{3}h[(y_0+y_n)+4(y_1+y_3+y_5\ldots)+2(y_2+y_4+\ldots)]$ or, in words, 'One third of h times (the sum of the first and last ordinates plus four times the sum of the odd ordinates plus twice the sum of the even ordinates)' where odd and even ordinates are those with odd and even suffixes respectively. Using Simpson's rule, the area under the curve shown in Fig. 2 would be:

$\frac{2}{3}[(4+8)+4(7+10+10)+2(9+11)]$.

Work out this area (to the nearest square unit).

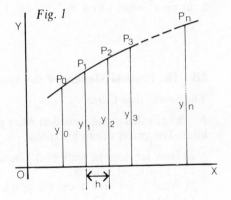

Fig. 1

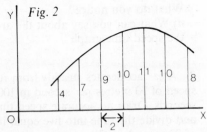

Fig. 2

Note Simpson's rule is derived by fitting a parabola to the three points $P_0 P_1 P_2$, and another parabola to the points $P_2 P_3 P_4$, etc. The derivation is a little beyond the scope of this book.

13 Use Simpson's rule to find the area under the curve drawn through the points whose co-ordinates are given:

    x 2 3 4 5 6 7 8
    y 0 5 8 9 8 5 0

14 Repeat question 13 for the following:

x	0	1	2	3	4	5	6	7	8
y	3	5	8	9	8	6	7	9	10

15 Use Simpson's rule to find the areas under the following curves:

a)

x	1	2	3	4	5	6	7	8	9	10	11
y	6	8	9	8	6	5	3	3	5	6	8

b)

x	5	7	9	11	13	15	17	19	21
y	0	3	5	7	8	8	6	4	1

c)

x	−2	1	4	7	10	13	16
y	1	1·6	2·0	2·2	2·6	3·4	4·5

16 Calculate the values of y for the given values of x. Use Simpson's rule to find the area under the curve in each case.

a) $y = x^2 + 2x - 4$ $x = 2, 3 \ldots 8$
b) $y = x^2 - x - 2$ $x = 2, 4, 6 \ldots 14$
c) $y = 2x^2 - 3x + 2$ $x = -3, 0, 3, 6, 9$

17 Repeat question 16 for the following:

a) $y = x^3$ $x = 0, 1, 2 \ldots 6$
b) $y = x^3 - 6x^2 + 6x + 10$ $x = 0, 1, 2 \ldots 6$

8B The Physical Meaning of the Area under a Curve

The Speed/Time Curve

1 Peter walks along a road at 6 km per hour. The graph shows his speed.

a) How far does he walk in i) 1 hour, ii) 2 hours, iii) 3 hours?
b) What is the area under the graph from $t = 0$ to i) $t = 1$, ii) $t = 2$, iii) $t = 3$?
c) Compare your answers to a and b. What do you notice?
d) What can you say about the area under the speed/time graph?

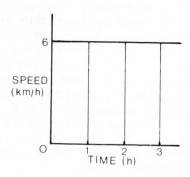

2 A car accelerates steadily from rest to a speed of 30 metres per second in 100 seconds. Draw an accurate speed/time graph and divide the time into five equal intervals of 20 seconds each.

a) Read off the average speed over each interval and hence calculate the distance travelled during that interval.

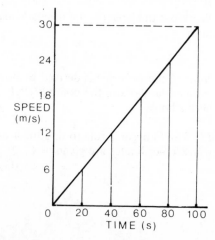

68

Record your results in a table like this:

Interval	1st	2nd	3rd	4th	5th
Average speed during interval	3 m/s	9 m/s			
Distance travelled during interval	3×20 $= 60$ m				

b) Calculate also the area under the graph in each of these five intervals (each area is a trapezium).

c) Comparing your answers to *a* and *b*, what do you notice?

d) Is the result you found in *1d* still true?

3 Repeat question 2 for a heavily loaded lorry on a motorway, accelerating steadily from a speed of 15 m/s to 21 m/s in one minute. This time divide the area under your graph into six equal intervals.

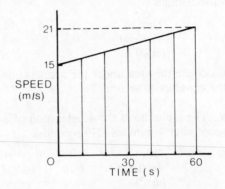

4 The table shows the speed of a car at the end of 5, 10...40 seconds after it moves off from rest.

Time (s)	0	5	10	15	20	25	30	35	40
Speed (m/s)	0	4·8	9·0	12·6	15·6	18·0	19·8	21·0	21·6

Draw a graph to illustrate these figures. Using Simpson's rule (or otherwise) calculate the approximate distance travelled during the first 40 seconds. Was it necessary to draw the graph?

5 Repeat question 4 for the following values:

Time (s)	0	2	4	6	8	10	12	14	16	18	20
Speed (m/s)	0	1	2·1	3·5	4·5	6·0	7·5	9·1	10·5	12·6	14·5

6 *a*) Water from a stream is flowing into a reservoir at a steady rate of 1000 cubic metres per hour. If time is plotted along the *x* axis and the rate of inflow along the *y* axis, what does the area under the graph represent?

b) There is a thunderstorm with heavy and prolonged rainfall. The flow of water is temporarily increased.

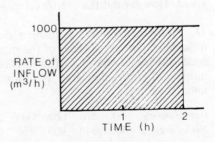

Time (h)	11 00	12 00	13 00	14 00	15 00	16 00	17 00	18 00	19 00	20 00	21 00	22 00	23 00
Rate of inflow (m³/h)	0	1500	1800	2050	2200	2150	2050	1900	1800	1700	1620	1550	1490

Calculate the approximate total volume of water that flows into the reservoir from this stream between 11 00 and 23 00 hours.

7 The diagram shows an acceleration/time graph, the acceleration being constant at $2 \, m/s^2$.

a) Starting from rest, what will the speed be after *i)* 1 sec, *ii)* 2 sec, *iii)* 4 sec?
b) What is the area under the 'curve' from $t=0$ to *i)* $t=1$, *ii)* $t=2$, *iii)* $t=4$?
c) What can you deduce about the area under the acceleration/time curve?

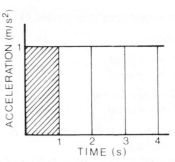

8 A car moves off from rest on a hilly road. The table shows its acceleration at various times.

Time (s)	0	5	10	15	20	25	30	35	40
Acceleration (m/s²)	0·5	0·6	0·7	0·7	0·55	0·35	0·20	0·10	0

Calculate the area under the acceleration/time curve. What is the approximate speed of the car after 40 seconds?

9 The table shows the acceleration of a train leaving a station and reaching a steady speed after 4 minutes (240 seconds).

Time (s)	0	30	60	90	120	150	180	210	240
Acceleration (m/s²)	0·20	0·19	0·17	0·14	0·10	0·06	0·03	0·01	0

a) Calculate the area under the acceleration time graph from $t=0$ to $t=240$.
b) What is the approximate final speed of the train?

Note In questions 10 to 13 the information is given at irregular intervals. To use Simpson's rule it is necessary to draw a graph and mark in ordinates dividing the area under the graph into an even number of strips of equal width.

∗10 The table shows the speed of a car at irregular intervals.

Time (s)	0	5	9	12	17	23	31	37	40
Speed (m/s)	0	3	6	8	11	14	17	19	20

Draw a graph and from your graph read off the speed at $t=0, 5, 10 \ldots 40$ seconds. Then using Simpson's rule (or otherwise) calculate the approximate area under the graph. How far did the car travel in 40 seconds?

11 Norman, the milk roundsman, delivers milk daily. On certain days of the month Jock, a market research assistant, checks the number of bottles delivered. Here are his figures:

Day of month	1st	5th	12th	16th	21st	24th
No. of bottles	860	840	900	860	910	880

Draw a graph to represent these results, showing the number of bottles delivered daily. Plot the first day at $d=\frac{1}{2}$, day 5 at $d=4\frac{1}{2}$, etc. Read off the number of bottles at $d=0, 4, 8 \ldots 24$.
By finding the area under the curve, calculate approximately the total number of bottles delivered in 24 days.

∗ 12 The table shows the number of units of electricity used by a householder on certain days during a spell of cold weather in February.

Day of month	1	5	8	11	15	18	20	23	25	28
Units used per day	6·4	9·1	9·5	8·8	7·8	7·9	8·7	10·5	11·6	12·0

Represent these figures on a graph, plotting the number of units used on the first day at $d=1$, on the 5th day at $d=5$, etc.
Read off the number of units at $d=0, 2, 4, 6 \ldots 28$ and hence, using Simpson's rule (or otherwise), calculate the approximate total number of units used during the 28 days of February in that year.

∗ 13 The table shows the rate at which a block of metal is cooling.

Time (minutes)	1st	3rd	6th	10th	15th	22nd
Rate of cooling (°C per min)	24·4	21·7	18·4	15·4	12·8	10·5

Show these results on a graph, this time plotting the rate of cooling for the first minute at $t=1$, for the 3rd minute at $t=3$, etc. Read the values of the ordinates at $t=0, 2, 4 \ldots 20$ minutes. Find the area under the curve from $t=0$ to $t=20$. What does this tell you?
If the initial temperature of the block was 550 °C, what is the approximate temperature after 20 minutes?

8C Miscellaneous

1 On graph paper taking 1 cm to 1 unit on both axes, with centre the origin, draw a quadrant of a circle of radius 10 cm. Calculate its area using the trapezium rule or Simpson's rule.
The correct answer should of course be $(\pi/4)(100)$ i.e. 78·5 cm². How close did you get?

2 Repeat question 1 with an ellipse. To draw an ellipse, mark axes in the centre of a sheet of graph paper with the x axis parallel to the long side of the paper. Insert drawing pins at $(6, 0)$ and $(-6, 0)$, tie a loose loop of strong cotton round the pins, insert the point of a pencil in the loop and draw the ellipse. With a little practice you will soon be proficient.
Using the trapezium rule or Simpson's rule, calculate the area of one quadrant of the ellipse. Your answer should be $(\pi/4)(ab)$. How close did you get?

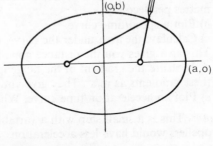

3 Repeat question 1 with a parabola. With centre $(4, 0)$ and radius 2, draw an arc and mark the point where it cuts $x=2$. Repeat with the same centre and radius 3, and mark it where it cuts $x=3$. Repeat with $x=4, 6, 8 \ldots 14$, marking where the arcs cut $x=4, 6, 8 \ldots 14$ respectively. Find the area under the curve. Your answer should be $\frac{2}{3}ab$.

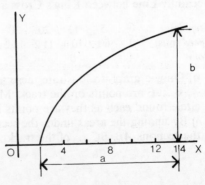

4 The table shows the distance travelled by a car at various times.

Time (s)	0	10	20	30	40	50	60
Distance (m)	0	300	550	750	900	1000	1050

Draw a distance/time graph and measure the gradients at $t=0, 10, 20\ldots60$ seconds. The gradient gives you the speed of the car.
Now draw a speed/time graph, and calculate the area under the curve. Your final answer should of course be 1050 metres, but there are bound to be drawing errors. How close did you get?

5 The table shows the temperature at various times of a quantity of water which is cooling slowly.

Time (min)	0	10	22	36	52	70	80
Temp (°C)	98	90	82	74	66	58	54

Plot a graph of temperature against time. Measure its gradient at $t=5, 15,\ldots75$ seconds. This gives you the rate of cooling.
Plot the rate of cooling against time. Find the area under the curve. Your answer should be $98°-54°=44°$. How close did you get?

✳ 8D Speed and Acceleration of Modern Transport

The questions in this section are based on information provided by courtesy of Three Quays Marine Service (P & O Group), London Transport, British Rail (Railway Technical Centre), British Leyland and the Clayton Dewandre Company.

1 The table shows the acceleration characteristics of a cruise ship of 8820 tonnes displacement, fitted with controllable pitch propellers.

Time in minutes	0·15	0·5	1·0	1·5	2·0	2·5	3·0	3·5	4·0	4·5	5·0
Speed in knots	0	3·2	7·7	10·5	12·1	13·0	13·7	14·1	14·4	14·6	14·8

a) Convert the speed to metres per second (1 knot=1 nautical mile per hour=0·514 metres per second).
b) Plot a speed/time curve.
c) Calculate the area under the curve, remembering to change your times to seconds. This area gives you the distance gone in the first five minutes.
d) Measure the gradient of the tangent to the curve at 0·5, 1, 2,…5 minutes. Express these gradients as m/s^2. They give you the accelerations.
e) Plot an acceleration/time curve. What is the maximum acceleration?

Note This is a small ship with a variable pitch propeller. A large ship with fixed pitch propellers would have less acceleration.

2 The table shows the speed of a London Underground train on the Eastbound Piccadilly Line between King's Cross and Caledonian Road stations.

Time (s)	0	5	13	20	30	40	50	65	80	100	120	145	150	158
Speed (m/s)	0	6·1	10·0	11·2	12·2	12·7	13·1	13·4	14·9	14·4	14·2	14·0	8·9	0
Point	A		B					C	D			E		F

a) Draw a graph to illustrate these figures.
A, B,…F are points on the track. Mark these points on your graph putting a circle round each as they are points in space, not times or distances.
b) By finding the areas under the sections AB, BC,… of the curve, find the lengths of the sections AB, BC,… of the track.

c) Sections *AB*, *CD* and *EF* are level and sections *BC* and *DE* slope up with a gradient of 1 in 85. Calculate the height of the track at Caledonian Road above the height of the track at King's Cross. (Answer to nearest metre.)

d) Describe the acceleration characteristics of each of the sections *AB*, *BC*,…

e) State the maximum acceleration and the maximum deceleration, and where they occur.

3 The table shows the speed of a Westbound train from Caledonian Road to King's Cross on the same line as in question 2.

Time (s)	0	5	10	20	30	40	62	66	110	128	136	146
Speed (m/s)	0	5·2	9·2	12·5	14·7	16·4	19·0	13·9	13·9	16·9	8·6	0
Point	G						H	I	J	K		L

a) Draw the curve and mark on the points *G* to *L* as in question 2*b*).

b) A speed restriction is in operation over the section of the line *I* to *J*. By calculating areas under the curve, find the distances from Caledonian Road at which this restriction starts and ends. Why do you think it is in force?

c) State the maximum acceleration and the maximum deceleration of the train and name the sections in which they occur.

4 The following table gives test values of the acceleration characteristics through the gears of the British Leyland Rover SD 1, 3500 cc.

Speed (mph)	0	30	40	50	60	70	80
Time (s)	0	3·1	4·5	6·4	8·6	11·4	15·4

a) Convert the speeds to metres per second (1 mph = 0·447 m/s).

b) Plot a speed/time graph.

c) Calculate the area under the curve. This gives the distance gone in 15·4 seconds.

d) Measure the gradient at 1, 3, 5,…15 seconds. These gradients give the acceleration.

e) Plot an acceleration/time curve. What is the maximum acceleration?

✳5 For the Rover in question 4, the measured time for $\frac{1}{4}$ mile from a standing start is 16·9 seconds and 30·8 seconds for 1 kilometre from a standing start. The maximum speed of the car is 126 mph.

a) Convert $\frac{1}{4}$ mile to metres and 126 mph to m/s.

b) Extrapolate your speed/time curve from question 4 by eye to 16·9 seconds and to 30·9 seconds. Calculate the areas under the extrapolated curve and see whether they come to $\frac{1}{4}$ mile and 1 km respectively. (It may help you to remember that you must not exceed the maximum speed, and to know that the time to accelerate from 70 to 90 mph in top gear is 10·3 seconds.) You cannot expect an exact answer as extrapolation is always prone to error.

6 Repeat question 4 for the Triumph TR 7, UK 2 Litre whose acceleration characteristics are shown in the table.

Speed (mph)	0	30	40	50	60	70
Time (s)	0	3·0	4·7	6·7	9·5	12·3

✳✳7 Repeat question 5 for the TR 7. The time, from a standing start, is 17·2 seconds for $\frac{1}{4}$ mile and 31·6 seconds for 1 km. The maximum speed is 114 mph and the time to accelerate from 60 to 80 mph in top gear is 9·0 seconds.

8 The table shows the speed/time characteristics of a high speed train on three different gradients. The middle one is representative of a high speed train leaving York on the Northbound route. (For railway enthusiasts, the locomotive is DDL DBMU 253 of 4500 BHP pulling a train of seven coaches, two of which are catering vehicles. The overall length of the train is 195 metres, and the overall mass 407 tonnes.)

Gradient	Time (min)	0	0·5	1·0	1·5	2·0	2·5			
2 in 100 down	Speed (mph)	0	35	65	85	103	118			

Gradient	Time (min)	0	1	2	3	4	5	6	7	8	9
level	Speed (mph)	0	45	71	87	98	107	114	120	125	128
2 in 100 up	Speed (mph)	0	25	43	51	55	58	59·5	60·5	61	—

a) Convert the above speeds to metres per second (1 mph = 0·447 m/s).
b) Draw a speed/time curve, plotting all three cases on one graph.
c) Measure the gradients of your curves at five or six points spaced evenly along the curves. These gradients are the accelerations in m/s per minute. Convert them to m/s^2 and plot three acceleration time curves, all on the same axes. Label the curves '2 in 100 down', 'level' and '2 in 100 up'.

9 Calculate the areas under the speed/time curves in question 8 at the following intervals: first curve (2 in 100 down) at intervals of 0·5 mins; other two curves at intervals of 1 minute. Remember to convert your time scale to seconds before calculating the areas. These areas give the distance gone. On a single graph plot the distance/time curves for the three gradients.

10

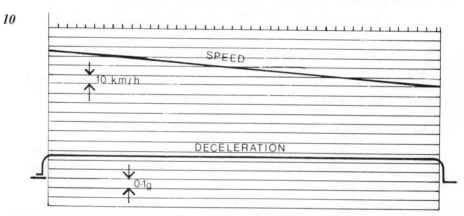

On 18th October 1976, tests on the front brakes of a family saloon car were carried out on a slippery surface (wet Bridport gravel). The speed of the front wheels (which can be taken as the speed of the car) was measured and recorded on a rotating drum. An accelerometer measured the deceleration of the car and this was recorded also on the same drum. The diagram shows a quarter scale reproduction of the relevant part of the paper taken from the drum after the test. The lines on the original were 1 cm apart and 1 cm represented a change of velocity of 10 km/h. For the deceleration, 1 cm represented a deceleration of 0·1 g, where g is the acceleration of gravity. As g is approximately 10 m/s^2, 1 cm represented a deceleration of 1·0 m/s^2. The marks along the top give the time scale, the interval between two successive marks being 0·1 seconds.

a) What is the length of time during which the deceleration lasted?
b) What was the deceleration in m/s^2?
c) What was the loss of speed during the test? Give your answer i) in km/h, ii) in m/s.
d) From your answers to c and a calculate the deceleration in m/s^2. Does it agree with your answer to b?

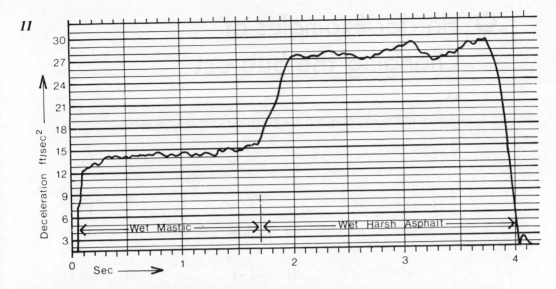

The curve shows the deceleration of a family saloon car with the brakes fully applied, running from a 'low coefficient' slippery surface on to a 'high coefficient' rough surface. Copy the curve into your book, smoothing out minor irregularities.

a) Find the area under the curve. This gives the loss of speed during the test in feet/second. Change this loss of speed to m/s (1 m = 3·28 ft).

b) Taking g as 10 m/s² find the average deceleration in terms of g i) on the smooth surface, ii) on the rough surface.

c) You would need one more item of information to calculate the distance travelled during the test. What information? If you had this information, how would you proceed? Would you expect to get an accurate answer?

9 Harder Examples of Changing the Subject

9A

In each of the following, write x in terms of the other letters:

1 $y = a + bx$

2 $y = b - cx$

3 $y = \dfrac{a}{bx}$

4 $\dfrac{a}{y} = \dfrac{b}{x}$

5 $y = 1 + \dfrac{1}{x}$

6 $y = \dfrac{1}{1+x}$

7 $y = ax^2$

8 $y = (bx)^2$

9 $c^2 y = x^2$

10 $y^2 = a^2 + x^2$

11 $y = \left(\dfrac{c}{x}\right)^2$

12 $x = ax + b$

13 $ax = bx + c$

14 $a(x + y) = bx$

15 $\dfrac{x}{1+x} = a$

16 $y = c\sqrt{x}$

17 $y = \sqrt{(x+a)}$

18 $y = \sqrt{(x^2 + b^2)}$

19 $y = c + \sqrt{x}$

20 $\dfrac{a}{\sqrt{x}} = \dfrac{b}{\sqrt{y}}$

9B

Rearrange each of the following formulae to make the given letter the subject:

1 $V = \pi r^2 h$ (r)

2 $V = \frac{1}{3} a^2 h$ (a)

3 $I = \dfrac{4}{3} ma^2$ (a)

4 $I = m(a^2 + d^2)$ (d)

5 $A = P + \dfrac{PRT}{100}$ (P) (R)

6 $F = \dfrac{9c}{5} + 32$ (c)

7 $V = w\sqrt{a^2 - x^2}$ (a)

8 $f = \dfrac{1}{2l}\sqrt{\dfrac{T}{m}}$ (T) (m)

9 $J = m\sqrt{2ga}$ (a)

10 $T = 2\pi\sqrt{\dfrac{a}{6g}}$ (a) (g)

9C

Rearrange each of the following to give x in terms of the other letters:

1 $x = \dfrac{x + a}{b}$

2 $a = \dfrac{y}{1+x}$

3 $a = 2b\sqrt{\dfrac{1}{x}}$

4 $y = a(b + \sqrt{x})$

5 $p = \sqrt{\dfrac{1-x}{1+x}}$

6 $a = \sqrt{\dfrac{b-x}{b+x}}$

7    $3r - 2(x+5) = rx$ 8 $m = \dfrac{2x}{3n-4x}$ 9 $y = \dfrac{3x+2a}{1-2x}$

10    $a = \dfrac{ax-y}{ax+y}$ 11 $5y = \sqrt{x} - 3$ 12 $y = \dfrac{x-z}{x}$

9D

1   If $p > q$ and $(p-q)^2 = 64$, express q in terms of p.

2   If $(m-n)^3 = 64$ express m in terms of n, and n in terms of m.

3   If $4a^2 + 9b^2 = 12ab$, write a in terms of b.

4   Given that $a(2x-1) = b(x-2)$, write x in terms of a and b.

5   Given that $\dfrac{p}{x} = \dfrac{q}{x+r}$, express x in terms of the other letters.
If $p = 16$, $q = 21$ and $r = 0.2$, calculate the value of x.

6   Given that $\sqrt{r^2 + s} = 2r$, find r in terms of s. Calculate the value of r when $s = 75$.

7   In the formula $y = Ax^2$, if $y = 54$ and $A = 1.5$ find the value of x. If $y = Bx^n$, express x in terms of B and y.

8   Given that $T = 62.5$ when $k = 1.6$ and $p = 2.5$, find the value of n in the formula $T = kp^n$.

9   If $x = at^2$ and $y = 2at$, find an equation which relates x and y, but does not contain t.

10   Given that $x = p^2 + 1$ and $y = 2ap$, find an equation which relates x and y.

11   A length l of fencing forms three sides of a rectangular enclosure. If x is the length of one of the equal sides and y is the length of the third side, write an expression

   a) for x in terms of l and y b) for y in terms of l and x.

12   A rectangular block on a square base of side b cm has a volume of V cm^3 and a height h cm. Write an expression for b in terms of V and h. If $V = 150$ cm^3 and $h = 24$ cm, find the value of b.

13   If $\dfrac{ax+2}{3} - \dfrac{bx}{2} = y$, find x in terms of the other letters.

14   If $2x = 5y$, find the value of $\dfrac{x+2y}{3x-y}$.

15   Given that $T\left(1 - \dfrac{a}{4}\right) = 1 + \dfrac{3a}{4}$, express a in terms of T.

16   $a - 2b = bc$. Express b in terms of a and c, and c in terms of a and b.

17 If $9-x=7$, and $a+xy=6$, find x and find the value of y when $a=11$.

18 If $T=\dfrac{5n+1}{7n-4}$, find T when $n=-\frac{1}{3}$. Express n in terms of T.

19 Rearrange the formula $\dfrac{1}{f}=\dfrac{1}{u}+\dfrac{1}{v}$ to make u the subject.

20 Rearrange the formula $m=\dfrac{v}{f}-1$ to make f the subject.

9E

1 The formula $a+(n-1)d$ gives the nth term in a series. If $a=2$ and $d=3$, find which term is 56.

2 The sum of the first n terms of a series similar to the series in question 1 is given by the formula $\dfrac{n}{2}[2a+(n-1)d]$. If $a=2$ and $d=3$, and the sum is 876, find n, the number of terms in the series.

3 The sum of a different type of series is given by 2^n-3, where n is the number of terms. Find the sum of the first five terms. If the sum is 125, how many terms are there?

4 A stone falls from the edge of a cliff. Its velocity v in metres per second is given by the formula $v^2=2gs$ where s is the distance it has fallen in metres. If $g=9.8\,\text{m/s}^2$, calculate how far it has dropped when its velocity is 12 m/s.
* Wind resistance reduces the square of the velocity by an amount which is proportional to the square of the distance dropped. Write a new equation for v^2, introducing a constant k. If when the stone has dropped the same distance as above its speed is 11·5 m/s and not 12 m/s, find the value of k.

5 The volume of a circular disc with a circular hole in the centre is given by the formula $V=a(R-r)(R+r)$ where a is constant and R and r are the outer and inner radii respectively. Rearrange the formula to make r the subject. Find the value of r when $V=10.5\,\text{cm}^3$, $a=1.75$ and $R=2.75\,\text{cm}$.

6 The formula for the area of a trapezium is $A=\frac{1}{2}(a+b)h$ where a and b are the lengths of the parallel sides and h is the perpendicular distance between them. Rearrange the formula to make a the subject.
The area of a trapezium is $31.625\,\text{cm}^2$, one of the parallel sides being of length 6·7 cm and the perpendicular distance between the parallel sides being 5·5 cm. Calculate the length of the other parallel side.

*7 $T=2\pi\sqrt{\dfrac{h^2+k^2}{gh}}$ is the formula giving the time in seconds of one oscillation of a compound pendulum. Rearrange the formula to make k the subject.
Taking g as $9.8\,\text{m/s}^2$, k and h are both distances measured in metres. If T is 2 seconds and $h=25\,\text{cm}$, calculate k correct to 2 significant figures.

8 R_t ohms is the resistance of some wire at $t\,^{\circ}C$, R_o ohms is the resistance at $0\,^{\circ}C$ and a is the temperature coefficient of resistance. Rearrange the formula $R_t = R_o(1 + at)$ to make a the subject. If R_o is 0·846 ohms and R_t is 0·85 ohms at $15\,^{\circ}C$, calculate the value of a.

9 The object distance u and the image distance v for a concave mirror of radius r, are connected by the formula $\dfrac{1}{v} + \dfrac{1}{u} = \dfrac{2}{r}$. Rearrange this formula to make v the subject. Find the value of v if $r = 34$ cm and $u = 51$ cm.

✳10 The formula $I = m\left(\dfrac{l^2}{12} + \dfrac{r^2}{4}\right)$ gives the moment of inertia of a cylindrical rod of mass m kg, length l and radius r, both in metres. Rearrange the formula to make r the subject. If $I = 0.028$ kg m^2, $m = 4.2$ kg and $l = 27$ cm, calculate the radius of the rod.

Miscellaneous Examples A

A1

1 $\mathcal{E} = \{11, 12, 13, 14, 15, 16, 17, 18\}$, $A = \{\text{Prime numbers}\}$, $B = \{x : x/3 \text{ is an integer}\}$, $C = \{x : 2x + 1 > 29\}$.
List the sets A, B, C, $A \cap C$, C', $(A \cup B \cup C)'$ and $B' \cap C$.
Draw a Venn diagram to show $\mathcal{E}$, A, B and C.

2 Find all the possible values between $0°$ and $360°$ for the angles A, B, C,.....I, given that:

$\sin A = 0.8480$	$\sin B = -0.9272$	$\cos C = -0.5592$
$\tan D = 1.3270$	$\cos E = 0.9100$	$\cos F = -0.7490$
$\tan G = -0.3000$	$\tan H = -2.9042$	$\sin I = -0.2250$

3 *a)* Write down the inverse of each of the following matrices. If there is no inverse, state the reason.

$i) \begin{pmatrix} 3 & 1 \\ 5 & 2 \end{pmatrix}$ $ii) \begin{pmatrix} 1 & 3 \\ 2 & 6 \end{pmatrix}$ $iii) \begin{pmatrix} -4 & -3 \\ 3 & 2 \end{pmatrix}$ $iv) \begin{pmatrix} 1 & -2 \\ 1 & 3 \end{pmatrix}$

b) Write down the inverse of $\begin{pmatrix} 2 & -3 \\ 1 & -2 \end{pmatrix}$ and use this to solve the equations
$\left. \begin{array}{l} 2x - 3y = 12 \\ x - 2y = 7 \end{array} \right\}$.

4 *a)* Given that $5p^2 + p^2 q = 512$, i) calculate p when $q = -3$,
ii) write an expression for q in terms of p.
b) If $a^2 b^2 - 2ab + 1 = 0$ write an expression for b in terms of a.

5 Given triangle ABC in which $AB = 16$ cm, $BC = 12$ cm and $AC = 9$ cm, calculate angle ACB.
AC is produced to D so that $CD = 6$ cm. Calculate the length of BD and angle CDB.

6 A solid metal cube which has edges of length 60 cm is melted down and reformed into two spheres the same size as one another. Find the radius of the spheres. Give your answer correct to 2 s.f.

A2

1 *a)* A and B are subsets of the universal set $\mathcal{E}$. If $n(A) = 7$, $n(B) = 17$ and $n(\mathcal{E}) = 30$, find the maximum and minimum values of $n(A \cap B)$.
b) If C and D are also subsets of the same universal set, and $n(C) = 13$ and $n(D) = 21$, find the maximum and minimum values of $n(C \cap D)'$.

2 *a)* If $\tan x° = -\frac{4}{3}$ and angle x lies between $180°$ and $360°$, write down $\sin x°$ and $\cos x°$.
b) Solve the following equation for values of x from $0°$ to $360°$: $2\cos^2 x - \cos x - 1 = 0$.
c) Complete the following:

$\sin 25° = \cos$____	or $\cos$____	
$\sin x° = \cos$____	or $\cos$____	
$\cos 140° = \sin$____	or $\sin$____	

3 *a)* Factorise $3x^2 + 7x - 6$ *b)* Solve $3x^2 + 7x - 6 = 0$
c) Solve $3x^2 + 7x - 1 = 0$ (Give your answers correct to 2 d.p.)

4 The diagram shows the motorway links between four towns A, B, C and D. The following matrix R gives the rail links between the same four towns:

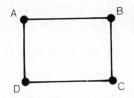

$$
\begin{array}{c}
\\
\\
From
\end{array}
\begin{array}{c}
\\
A \\ B \\ C \\ D
\end{array}
\overset{\begin{array}{c} To \\ A\ \ B\ \ C\ \ D \end{array}}{
\begin{pmatrix}
0 & 1 & 0 & 1 \\
1 & 0 & 1 & 0 \\
0 & 1 & 0 & 0 \\
1 & 1 & 0 & 0
\end{pmatrix}}
$$

a) Write the direct route matrix M, giving the motorway links. Copy the diagram and add to it the rail links.

b) Work out i) $M+R$ ii) R^2 iii) MR.

Describe what each of these matrices is showing.

5 Complete the table for the set $S = \{1, 3, 5, 7\}$ (mod 8) under the operation of multiplication. Use your table to answer these questions:

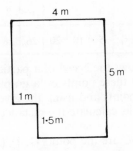

×	1	3	5	7
1	1			
3		1	7	
5				
7				

a) What is the identity element of this set under this operation?

b) Write down the inverse of each element of S.

c) Show that $(3 \times 5) \times 7 = 3 \times (5 \times 7)$.

d) If $y \in S$, find the values of y in each of the following:

i) $3y = 5$ (mod 8)   ii) $y/5 = 3$ iii) $y \times (y \times y) = y$

6 The diagram shows the shape and dimensions of the floor of a room which is 2·5 m high. The door and windows occupy 12 m². Find:

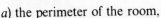

a) the perimeter of the room,

b) the area of the walls excluding doors and windows,

c) the cost of papering the walls and ceiling at the rate of £1·50 per m²,

d) the cost of carpet at £7·80 per m²,

e) the volume of air in the room.

If the heating system changes the air at the rate of 10 m³ every half-hour, find how long it takes to change the air completely. Give your answer to the nearest minute.

A3

1 P, Q and R are subsets of the universal set $\mathscr{E}$. Given that $n(\mathscr{E}) = 60$, $n(P \cap Q \cap R) = 7$, $n(P' \cap Q' \cap R') = 12$, $n(R) = 16$, $n(Q) = 21$, $n(R \cap Q) = 9$, $n(P \cap Q) = 15$ and $n(P \cup Q) = 44$, calculate:

a) $n(P' \cap Q \cap R')$ b) $n(P \cap Q' \cap R')$ c) $n(P' \cap Q' \cap R)$

2 A rectangular frame x cm by $(x+6)$ cm holds a mounted picture leaving a border of 2 cm all round the picture. Write an expression for the area of the picture. If the area of the picture is 160 cm², find the dimensions of the frame.

3 A ship is sighted on a bearing of 028° from coastguard station A. From station B, which is 6 km due east of A, the ship is on a bearing of 315°. Find the distance of the ship from A.

If the ship is drifting at 4 km/h on a course of 045°, find its distance and bearing from B one hour later.

4 Complete the incidence matrices for the given network.

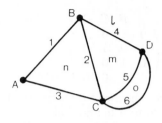

$$P = \begin{array}{c} A \\ B \\ C \\ D \end{array} \begin{pmatrix} 1 & 2 & 3 & 4 & 5 & 6 \\ 1 & 0 & 1 & 0 & 0 & 0 \\ & & & & & \\ & & & & & \\ & & & & & \end{pmatrix}$$

$$Q = \begin{array}{c} 1 \\ 2 \\ 3 \\ 4 \\ 5 \\ 6 \end{array} \begin{pmatrix} l & m & n & o \\ 1 & 0 & 1 & 0 \\ & & & \\ & & & \\ & & & \\ & & & \\ & & & \end{pmatrix} \qquad R = \begin{array}{c} A \\ B \\ C \\ D \end{array} \begin{pmatrix} l & m & n & o \\ 1 & 0 & 1 & 0 \\ & & & \\ & & & \\ & & & \end{pmatrix}$$

a) Find the matrix product PQ.
b) What do you notice about PQ and R?
c) Write down the matrix P' showing the incidence of arcs on nodes.
d) Find the product PP'. How does this matrix relate to the direct route matrix for the network?

5

Time (s)	0	2	4	6	8	10	12	14	16
Speed (m/s)	5	10	20	26	28	23	16	12	10

The table shows the speed of a particle moving in a straight line for a period of 16 seconds. Plot these points on a graph using 2 cm to 2 seconds on the horizontal axis. Join up the points and using the given points only and either the trapezium rule or Simpson's rule, calculate the distance travelled by the particle in the 16 seconds.

6 A, B and C are the points $(1, 1)$, $(3, 1)$ and $(1, 4)$ respectively. M is the matrix $\begin{pmatrix} 1 & -1 \\ 1 & 1 \end{pmatrix}$. Find the images A', B', C' when the position vectors of A, B, C are premultiplied by M. Show the triangles ABC and $A'B'C'$ on the same diagram and describe the transformation represented by M.

The matrix N is $\begin{pmatrix} 0 & -1 \\ -1 & 0 \end{pmatrix}$. $A''B''C''$ is the image of $A'B'C'$ when it is operated on by the matrix N. Show this third triangle on your diagram.
Write down the matrix M^{-1}. State the single geometrical transformation which is represented by $M^{-1}NM$.

A4

1 a) Find the values of a and b if $\begin{pmatrix} 5 & 2 \\ 7 & 3 \end{pmatrix}\begin{pmatrix} 1 \\ -2 \end{pmatrix} = \begin{pmatrix} a \\ b \end{pmatrix}$.

b) Find the matrix M if $M\begin{pmatrix}1\\1\end{pmatrix}=\begin{pmatrix}1\\-2\end{pmatrix}$.

c) Solve the simultaneous equations $\left.\begin{aligned}5x&=4-2y\\3y&=5-7x\end{aligned}\right\}$.

2   The volume of a cylinder of radius r and length l is $\pi r^2 l$. If a cylindrical hole of radius $r/2$ is bored through the whole length of the cylinder, write down the volume removed. Hence write down an expression for the volume remaining.
If the volume of metal in such a 'pipe' of length $2\,m$ is $2169\,cm^3$, find the outer radius.

3   The table gives the depth of a river bed in metres, measured at intervals of $2\,m$ from one bank.

Distance from bank	0	2	4	6	8	10	12	14	16
Depth	0	0·4	1·0	2·1	3·4	3·8	3·0	1·5	0

By plotting these values on a graph, draw a section across the river. If the normal water level is $1\,m$ below the level of the bank, find the cross-sectional area of the water, using the plotted points only. The water flows at a mean rate of $0·5\,m/s$. Find the volume of water passing over a weir in the river in one minute.
If after prolonged rainfall the level of the water rises $0·6\,m$ and the speed of flow increases to $0·8\,m/s$, find the increase in the volume of water flowing over the weir in one minute.

4   On the same axes draw the graphs of $y=2\cos x°$ and $y=\cos 2x°$. Take $0\leqslant x\leqslant 360$. Use your graph to solve the equation $\cos 2x-2\cos x=0$.

5   The diagrams below show the plan and elevation (not to scale) of a greenhouse, ignoring details of door, windows, etc. The plan form is a regular octagon.

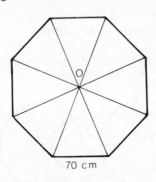

70 cm

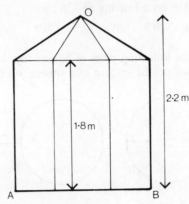

2·2 m

1·8 m

a) Make a careful scale drawing of the plan and elevation.
b) On the plan mark a length XY which is equal to the length AB on the elevation. Give the value of XY correct to 2 s.f.
c) Calculate the area of ground covered by the greenhouse.
d) By treating the greenhouse as a prism capped by a pyramid, calculate its volume in cubic metres to 2 s.f.
e) Make a careful scale drawing of one actual triangular face of the roof. What is its height? Calculate its area.
f) What is the angle between one triangular face and the horizontal?
g) What is the total area of glass in the greenhouse?

6 *a*) Copy and complete the following table and draw the graph of $y=4/x$ for the domain $0\cdot25-4$.

x	0·25	0·5	0·75	1	1·5	2	2·5	3	3·5	4
y			5·3						1·1	

b) A field is to be laid out in the form of a rectangle with sides x and y (both measured in units of 100 metres). Its area is to be not less than 4 hectares. Write an inequality concerning its area. It is to be fenced all round, but only 2200 metres of fencing are available. Write another inequality concerning the length of the fence. If it is decided to make the length y at least twice as great as the width x, write a third inequality in x and y.

c) Add two lines to the graph you have already drawn to show the region in which x and y must lie.

d) If the width is to be as large as possible, give the dimensions of the field (to the nearest ten metres) and state its area.

e) Find the limits between which the length of the field must lie.

A5

1 Solve *a*) $\dfrac{3}{x-1}=1+\dfrac{4}{x+5}$ *b*) $\dfrac{2x}{x^2-4}=1+\dfrac{1}{x+2}$

2 A field bounded by four straight hedgerows was surveyed from corner A. The measurements taken from A to the other corners were:

    AB, 105 m on a bearing of $067°$;
    AC, 98 m on a bearing of $042°$;
    AD, 71 m on a bearing of $326°$.

Calculate *a*) BC *b*) CD *c*) BD *d*) the angle between BC and CD.

3 Are the following pairs of figures topologically equivalent *i*) if drawn on a plane, *ii*) if drawn on the surface of a sphere, *iii*) if made of thin flexible wire?

a)   *b*)

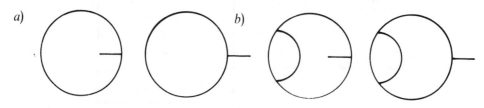

c)

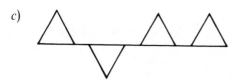
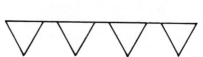

d)

4 Find the solution of the equation $(x-2)(x-4)=0$,

   a) in ordinary arithmetic   *b)* in mod 7 arithmetic   *c)* in mod 6 arithmetic

   d) in mod 12 arithmetic   *e)* in mod 24 arithmetic.

5 Braunston Tunnel is one of the longest canal tunnels on the Inland Waterways System in Great Britain. It is nearly 200 years old and the brickwork had decayed extensively, so in December 1978 the work of renewing the brickwork above the waterline was begun. The first diagram shows the approximate dimensions of the tunnel. For simplicity the sides are shown straight but they are actually curves on a radius of just over 6 metres. The second diagram shows the nominal dimensions of one 'engineering brick'.

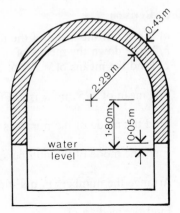

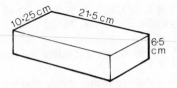

a) Adding 1 cm to each dimension of the brick to allow for the thickness of the mortar, find the volume of one brick in cubic metres.

b) The length of the tunnel is 1867 m. Find the total volume to be 'bricked up' (shown shaded in the diagram). Give your answer in cubic metres in standard form correct to 2 s.f.

c) How many bricks will be required altogether? Give your answer in standard form correct to 2 s.f.

d) If the weight of one brick is approximately 2·9 kg, what is the total weight of bricks in tonnes correct to 2 s.f.?

e) If it takes 20 000 bricks to build a pair of semi-detached houses of a certain design, how many such houses could be built with the bricks used in the tunnel?

(*Information by courtesy of British Waterways and Messrs Miller Buckley Civil Engineering Ltd of Rugby.*)

6 The parallelogram P has vertices A $(-1, 0)$, B $(1, 0)$, C $(3, 2)$ and D $(1, 2)$.

Operate on it with the matrix M_1 $\begin{pmatrix} 1 & -1 \\ 0 & 1 \end{pmatrix}$ to obtain the image $A'B'C'D'$.

Call this image P_1. Draw a diagram to show P and P_1 and describe the transformation represented by M_1.

Operate on P_1 with the matrix M_2 $\begin{pmatrix} -1 & 0 \\ 0 & -1 \end{pmatrix}$ to obtain the image $A''B''C''D''$.

Call this image P_2. Show this image on the same diagram as above. Describe the transformation represented by M_2.

If P_1 is reflected in the y axis and P_2 in the x axis, the two images are identical. Call them P_3. Write down the matrices which represent these reflections, and hence find the matrix which transforms P to P_3 directly. Confirm by actual multiplication that the matrix is the same for both series of operations.

10 Statistics

10A Revision

1 a) Work out the mean of the numbers: 6, 7, 8, 10, 15.
   b) Write down the means of: i) 60, 70, 80, 100, 150 ii) 0·6, 0·7, 0·8, 1, 1·5.
   c) Find the means of: i) 2·6, 2·8, 3·2, 2·7, 2·4 ii) 296, 297, 297, 295, 294.

2 The mean of 2, 5 and x is 4. Find x.

3 Find the mode of the numbers: 6, 6, 7, 8, 9.

4 Find the mode of the numbers: 6, 2, 8, 2, 9.

5 Arrange the numbers in question 4 in numerical order and write down the median.

6 Find the median of:

   a) 2, 3, 5, 6, 7, 8, 9, 9, 10 b) 3, 4, 4, 2, 5, 3, 1, 0, 6 c) 2, 4, 6, 8
   d) 24, 25, 26, 27, 28, 29 e) 3, 3, 3, 4, 5, 5, 5, 5, 5, 6 f) 3·4, 2·6, 4·5, 3·6

7 Write down the range for the numbers in each part of question 6.

8 Find the mean, mode and median of: a) 8, 7, 3, 3, 6 b) 5, 7, 7, 11.

9 The mean of 4 numbers is 24 and the mean of 5 other numbers is 60. Find the mean of the combined group of nine numbers.

10 a) If 5 numbers are arranged in numerical order, which number is the median?
    b) If 15 numbers are in numerical order, which is the median?
    c) If 4 numbers are in numerical order, how would you find the median?
    d) Which is the median of 12 numbers?
    e) If there are n numbers, which is the median i) if n is odd, ii) if n is even?

11 Each member of a form was asked how many children there were in their family.
Here are the results: 3 4 2 2 2 2 2 2
                       2 1 1 3 2 3 3 1
                       1 2 2 1 4 5 1
                       2 2 3 2 6 4 2

Copy and complete the following table which arranges the information in groups:

No. of children per family	1	2	3	4	5	6
No. of families (frequency)	6					

Draw a frequency diagram to display the data, being careful to label both axes. Copy and complete the second table and calculate the mean number of children per family.

No. of children per family (c)	No. of families (f)	Total number of children ($c \times f$)
1	6	6
2		
...		
6		
Totals:	30	

12 On a particular Saturday, the goals scored in 40 football matches were as follows:

No. of goals:	0	1	2	3	4	5	6
No. of matches:	5	13	4	10	5	2	1

Draw a frequency diagram to illustrate the data.
Draw up a table similar to that in question 11, and calculate the mean number of goals scored per match.

10B Grouped Frequencies: Histograms

1 In a science laboratory a class were given lengths of metal rod with carefully squared ends and told to measure their lengths to the nearest tenth of a centimetre, using a centimetre scale. Here are the results:

$$6\cdot9 \quad 5\cdot5 \quad 7\cdot8 \quad 8\cdot2 \quad 4\cdot5 \quad 7\cdot1$$
$$5\cdot1 \quad 7\cdot9 \quad 6\cdot2 \quad 5\cdot4 \quad 6\cdot4 \quad 8\cdot3$$
$$9\cdot3 \quad 4\cdot6 \quad 4\cdot9 \quad 7\cdot5 \quad 5\cdot9 \quad 7\cdot3$$
$$8\cdot6 \quad 8\cdot1 \quad 5\cdot8 \quad 7\cdot5 \quad 6\cdot7 \quad 7\cdot3$$
$$7\cdot7 \quad 9\cdot0 \quad 5\cdot9 \quad 7\cdot8 \quad 8\cdot5 \quad 6\cdot5$$

Group the data and complete the frequency table below. Fives are rounded upwards, i.e. 6·5, 6·6, etc., count as 7, but 6·4, 6·3 count as 6.

Length (to nearest cm)	5	6	7	8	9
Frequency	5				

Draw a frequency diagram to show the distribution. Your first block should extend from 4·5 to 5·5 and the others similarly.

2 Using a travelling microscope, the lengths of the metal rods in question 1 were measured again, this time to the nearest 0·01 cm. Here are the results:

$$6\cdot93 \quad 5\cdot14 \quad 9\cdot35 \quad 8\cdot63 \quad 7\cdot74 \quad 5\cdot53 \quad 7\cdot90 \quad 4\cdot58 \quad 8\cdot09 \quad 9\cdot01$$
$$7\cdot83 \quad 6\cdot17 \quad 4\cdot94 \quad 5\cdot85 \quad 5\cdot92 \quad 8\cdot23 \quad 5\cdot43 \quad 7\cdot55 \quad 7\cdot49 \quad 7\cdot76$$
$$4\cdot53 \quad 6\cdot40 \quad 5\cdot91 \quad 6\cdot74 \quad 8\cdot53 \quad 7\cdot13 \quad 8\cdot27 \quad 7\cdot31 \quad 7\cdot27 \quad 6\cdot52$$

Group the data, and complete the table below. Lengths on the boundary between two groups go into the upper group.

Length (cm) to 2 s.f.	4·5–5·0	5·0–5·5	....	9·0–9·5
Frequency		3		

Draw a frequency diagram to illustrate this data.

3 The eggs on a poultry farm are collected, washed, graded and packed ready for despatch. Overleaf is a list of the weights of 40 eggs selected at random before grading. Weights are given to the nearest gram.

61 72 67 51 55 74 63 65 76 62 73 69 71 72 87 80 57 68 54 64
71 65 81 62 75 73 66 62 71 67 57 76 84 75 66 59 74 64 68 79

Copy and complete the following table:

Weight (g)	50–55	55–60	....	85–90
Frequency		2		

Weights on the boundary between two groups go into the higher group. Draw a frequency diagram to illustrate the distribution.

4 Here are the distances travelled to school each morning by the members of a class of 30. All distances are rounded off to the nearest 0·1 km. Arrange them in groups 0–0·5, 0·5–1·0, ... 2·3–3·0, over 3·0. In cases of doubt, put distances into the higher group. Illustrate your results with a frequency diagram.

1·1 0·1 0·3 0·2 1·3 0·1 2·2 0·2 3·1 0·3 4·6 0·4 0·8 0·3 2·1
0·2 1·4 0·6 0·3 2·5 0·2 0·4 0·9 0·2 0·7 1·5 0·1 0·4 2·7 0·1

5 The table shows the number of hours of sunshine per day at a seaside resort during the month of May.

11·5 8·0 4·3 10·8 8·1 2·1 11·4 9·4
3·8 7·1 8·4 5·1 12·5 10·0 9·6 8·2
8·3 4·9 6·2 11·6 9·1 3·2 9·1 1·1
7·9 10·7 3·7 8·5 9·8 10·4 2·7

Draw up a frequency table using the groups 0–2, 2–4, ... 10–12, over 12.   2, 4, 6, etc., go in the higher group. Draw a frequency diagram using these grouped frequencies.

6 The table shows the distribution of the prices of houses for sale in a certain area one weekend.

Prices in thousands of pounds	3–10	10–15	15–20	20–30	30–40
Frequency	14	15	16	24	4

A house in the group 3–10 costs £3000 or more but less than £10 000. Represent the data on a frequency diagram.

Histograms

7 In a frequency diagram, as we have seen, the height of each 'block' represents the frequency. In a *histogram*, the *area* of each block represents the frequency. Use the data of question 6 to answer the following questions:

a) The group 3–10 represents prices between 3 and 10 (thousands of pounds). This is a difference of 7. What would be the height of the block if the width is 7 and the area represents the frequency, 14?

b) The group 10–15 has a width of 5. What is the frequency of this group? What should its height be?

c) Explain why the height of the group 15–20 must be 3·2. Find the heights of the other two groups. Draw the histogram. Label the vertical axis 'frequency per £1000'.

8 The marks obtained by 200 candidates in an examination are given below. There are no half marks.

Mark	0–9	10–19	20–29	30–39	40–49	50–59	60–69	70–79	80–89	90–100
Frequency	0	2	4	8	38	60	47	25	14	2

Draw a histogram to illustrate this data. Although there are no half marks, the group 10–19 must be drawn from 9·5 to 19·5, etc. Why?

9 The following table shows the distribution of the areas of 50 advertisements in a local newspaper:

Area (cm^2)	Frequency
under 200	24
200 and under 400	8
400 „ „ 600	4
600 „ „ 800	3
800 „ „ 1000	4
1000 „ „ 2000	3
2000 „ „ 3000	4

Illustrate this data by means of a histogram.

10 A girl sold cups of coffee at 10p a cup to raise money for charity. Her takings included exactly thirty 10p pieces. She noted their dates of issue as:

1970	1976	1968	1965	1969	1969	1968	1969	1971	1973
1968	1969	1968	1948	1950	1968	1968	1974	1975	1968
1969	1976	1969	1976	1968	1963	1950	1971	1973	1976

a) If she chose a coin at random, what is the probability that it would be issued *i)* in 1968, *ii)* between 1970 and 1973 inclusive?
b) Draw up a frequency table, grouping the data as follows:

Year of issue	Frequency
1945–1949 incl.	
1950–1954 „	
etc.	

Draw a histogram to illustrate this data. Between what numbers on the scale will the years 1945–1949 extend?
c) Using the block of your histogram which represents the years 1965–69, draw dotted lines parallel to the vertical axis to show the beginning and end of the year 1968. Find the area of this strip. What does this area represent? Divide this number by 30 to find the probability of a coin, chosen at random from the takings, being issued in 1968.
d) Draw a line on the histogram to mark the end of the year 1973. Find the area of the block representing the years 1970–73 inclusive. Hence, find the probability of a coin being issued between 1970 and 1973.
e) Why do your answers in *c* and *d* disagree with those in *a*?

11 A boy wrote down the months in which his classmates were born. The results are shown below:

Month	Jan	Feb	Mar	Apr	May	June	July	Aug	Sept	Oct	Nov	Dec
Frequency	3	3	3	4	1	4	0	1	5	2	1	3

a) Are the results surprising?
b) The boy grouped the data into the four quarters of the year and then drew a histogram. Draw the histogram and criticise the result.
c) What is the probability of someone chosen at random from the class being born in May? Give your answer *i)* using the original data *ii)* from the histogram.

12 After the first 14 matches of the season, each of the 80 teams in Divisions *A*, *B* and *C* of a league had scored between 5 and 39 goals, as recorded below:

No. of goals	5–9	10–14	15–19	20–24	25–29	30–34	35–39
Frequency	2	11	28	23	11	3	2

Draw a histogram to display the data. Your first block must extend from 4·5 to 9·5 goals, and the others similarly. Use your histogram to estimate the probability of a team scoring 15 goals.

13 Using the data given in question 5, arrange the frequencies in the following groups: 1, 2–3, 4–7, 8, 9–10, 11–12, where 2–3 include 1·5 to 3·4, etc., the fives being rounded upwards. Draw a histogram to illustrate these frequencies. From your histogram,

   a) state the probability of 11 hours of sunshine (to the nearest hour) on a given day in that May.
   b) state the probability of 5 hours of sunshine (to the nearest hour) on a given day in that May.
   c) Calculate also these probabilities from the original figures in question 5. Do they agree with probabilities found from the histogram? If not, say why not.

14 At an annual show and fête there was a prize for the dog with the longest tail. Entries were restricted to dogs less than 60 cm tall. Here is a summary of the judge's measurements:

Length of tail (cm)	20–22	23–24	25	26	27–28	29–31
Frequency	3	17	28	23	5	4

Copy the table into your book, remembering that the group 23–24 means 22·5 to just under 24·5, etc. Add another row to the table giving the width of each block. Draw a histogram.

15 *a)* Draw i) a frequency diagram, ii) a histogram to show the frequencies of the heights of children shown in the following table. (Heights on the boundary between two groups were included in the higher group.)

Height (cm)	120–125	125–130	130–135	135–140	140–145	145–150
Frequency	3	5	8	7	4	3

Plot values from 120 to 150 along the horizontal axis.
b) Draw i) a frequency diagram, ii) a histogram to show the frequencies of the heights of children given in the following table.

Height (cm)	120–126	126–132	132–134	134–137	137–142	142–150
Frequency	4	6	6	8	4	2

   c) What is the difference between a histogram and a frequency diagram?

✱ *16* Why does a grouped distribution of a set of data often give a clearer picture of the data than an ungrouped distribution? What other advantage is there in grouping data? Are there any disadvantages?

10C Finding the Mean of a Grouped Distribution

1 Using the following data, find the mean number of goals scored in a match:-

No. of goals	0	1	2	3	4	5
No. of matches	6	13	7	10	3	1

2 The number of letters in the names (surname plus one forename) of the first year's intake of 120 pupils in a school was as follows:

No. of letters	6	8	9	10	11	12	13	14	15	16	18
Frequency	1	2	7	14	10	25	28	17	8	6	2

Using a working mean of 12, copy and complete the following table:

Number of letters (n)	Frequency (f)	Deviation from working mean (d)	Total deviation (f × d)
6	1	−6	−6
8	2	−4	−8
...	...	...	...
18		+6	

Sum the last column and divide by 120. This gives the 'mean deviation from the working mean'. What is the true mean?

3 Repeat question 2 using a working mean of 14. Does the final value of the true mean vary with the choice of working mean?

4 The data used in 10B question 12 is repeated below. We wish to find the mean number of goals scored per match.

No. of goals	5–9	10–14	15–19	20–24	25–29	30–34	35–39
Frequency	2	11	28	23	11	3	2

As we do not know exactly how many goals were scored by each team, we use the middle value of each group as an average for the group. This is called the 'mid-interval value' (MIV). Use a working mean of 17, copy and complete the following table, and hence find *approximately* the mean number of goals scored by a team. Why *approximately*?

No. of goals	Frequency (f)	MIV	Deviation of MIV from working mean (d)	Total deviation (f × d)
5–9	2	7	−10	−20
10–14	11	12	−5	−55
.....	...	...	...	...
35–39	2			

5 Using the data of 10B question 9, taking the MIV's as 100, 300, 500, etc., and choosing a working mean of 500, estimate the mean area.

6 The number of pages in the books on a library shelf were grouped as follows:

No. of pages per book	0–100	101–200	201–300	301–400	401–500	501–600	601–700
Frequency (no. of books)	2	13	5	8	7	4	1

Taking the MIV's as 50, 150, 250, etc., estimate the mean number of pages per book.

7 A survey of house prices in an area produced the following data:

Price (£)	Frequency	Price (£)	Frequency	Price (£)	Frequency
5000–6000	3	14 000–16 000	11	24 000–26 000	2
6000–8000	15	16 000–18 000	3	26 000–28 000	1
8000–10 000	1	18 000–20 000	6	28 000–30 000	1
10 000–12 000	0	20 000–22 000	1	30 000–40 000	3
12 000–14000	4	22 000–24 000	3		

Houses costing £6000, £8000, etc., were included in the higher group. Estimate the mean price.

8 The marks obtained by 200 candidates in an examination are given below:

Mark	0–10	11–20	21–30	31–40	41–50	51–60	61–70	71–80	81–90	91–100
Frequency	0	2	4	8	38	60	47	25	14	2

Estimate the mean mark.

9 Using the data given in 10B 14, find the mean length of a dog's tail. Use a working mean of 25 cm.

10 Find the mean height of the children in the second table in question 10B 15. Choose your own working mean.

10D Cumulative Frequency Graphs

Discrete Distributions

1 Copy and complete the following table which shows the examination marks of 30 candidates:

Mark	Frequency	Cumulative frequency	Meaning of column 3
0–5	1	1	1 candidate had 5 marks or less
6–10	2	3	3 candidates had 10 marks or less
11–15	3		
16–20	8		
21–25	10		
26–30	6		

2 *a*) Draw a cumulative frequency graph for the data in question 1. Plot marks (*m*) along the 'horizontal' axis and cumulative frequencies (*f*) along the 'vertical' axis. (The curve is also known as an *ogive*.)
b) Using your graph, state how many candidates had *i*) 22 marks or less, *ii*) 25 marks or less, *iii*) 17 marks or less. Round fractional answers downwards to whole numbers. Why?
c) Which of the answers in *b*) are accurate and which are 'estimates'? Why?
d) Add dotted lines to your graph, as in the diagram, to show the median mark. (If the median mark is *x*, half the candidates have *x* marks or less.) What is the median mark?

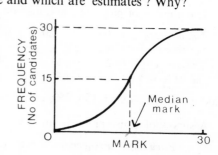

3 The table shows the marks of 100 candidates in a mathematics examination. There were no half marks. Calculate the cumulative frequencies and draw a cumulative frequency graph.

Mark	0–10	11–20	21–30	31–40	41–50	51–60	61–70	71–80	81–90	91–100
Frequency	0	1	3	6	9	17	28	29	4	3

Using your graph, answer the following questions.

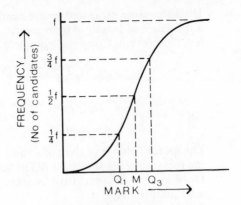

a) What is the median mark M?

b) What is the lower quartile Q_1 (one quarter of the pupils will have a mark of Q_1 or less)?

c) What is the upper quartile Q_3 (three-quarters of the pupils will have a mark of Q_3 or less)?

d) What is the interquartile range, Q_3-Q_1?

e) Why is the interquartile range sometimes more useful than the full range?

Note The median and quartile marks are used for processing marks (see 10F 6). They should be left as decimals.

4 Using the data of 10C 4, draw a curve of cumulative frequency and find the median number of goals scored and the interquartile range.

Note Although fractions of goals cannot be scored, the median and quartiles may come out as fractions or decimals. They should not be rounded off to whole numbers.

5 Using the data of 10C 6 draw a curve of cumulative frequency and find the median number of pages in a book and the interquartile range. The librarian notices that people do not like 'fat' books. Very few books containing more than about 450 pages are selected. Assuming the books on this shelf are typical of the whole library, what percentage of the books in the library would rarely be issued?

6 The marks obtained by 80 pupils in a test were as shown:

Mark	0–10	11–20	21–30	31–40	41–50
Frequency	3	13	28	26	10

Draw a curve of cumulative frequency and estimate

a) the percentage of pupils who scored over 45 marks

b) the percentage of pupils who failed, the pass mark being 20.

Continuous Distributions

In plotting cumulative frequency curves for continuous distributions it is necessary to think carefully what is the highest point of each group, and to plot the cumulative frequency at that point. There is no hard and fast rule. Each case must be considered on its merits.

7 A class of 32 pupils measured the difference in length of the diagonals of a tennis court to see if the court was truly rectangular. They worked in pairs, each holding one end of a long tape. Each pair carried out the measurements twice and all the 32 results were pooled. This is what they found:

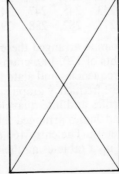

Difference in length (cm)	0	1	2	3	4	5	6	7	8
Frequency		1	1	3	4	6	7	5	3	2

Distances were measured to the nearest cm and fives were rounded downwards, so that a difference of length of 6 cm means a difference greater than 5·5 cm but equal to or less than 6·5 cm. Copy and complete the following table:

Difference (cm)	Frequency	Cumulative frequency	Plotting point
0	1	1	0·5
1	1	2	1·5
2	3	5	2·5

The fourth column of the table shows the plotting points. Thus in the third line, 5 of the measurements showed a difference of 2·5 cm or less.
Draw a curve of cumulative frequency and calculate the median, quartiles and interquartile range.
Was the tennis court truly rectangular?

8 Using the homemade apparatus shown, a class measured the lengths of 60 potatoes chosen at random from a 25 kg bag.

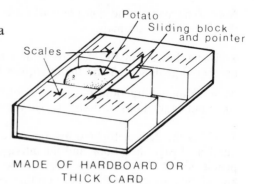

MADE OF HARDBOARD OR THICK CARD

Here are their results:

Length (mm)	50–60	60–70	70–80	80–90	90–100	100–110	110–120	120–130
Frequency	1	4	7	12	13	11	8	4

This time lengths of 60 mm, 70 mm, etc., are included in the *lower* group.
a) Draw up a table showing the cumulative frequencies and the plotting points.
b) Draw the curve of cumulative frequency and estimate the median, the quartiles and the interquartile range.
c) If lengths of 60 mm, 70 mm, etc., had been included in the *upper* group, where would the plotting points have been?

9 The table shows the weights in grams of 76 potatoes chosen at random from a 25 kg bag.

114, 124, 131, 138, 142, 147, 151, 152, 158, 161, 163, 163, 167, 169, 170, 172, 172, 175, 178, 181, 182, 182, 184, 184, 186, 188, 189, 191, 198, 198, 199, 201, 203, 203, 203, 204, 209, 210, 211, 211, 212, 214, 216, 219, 219, 220, 221, 222, 223, 224, 229, 230, 234, 235, 235, 238, 239, 241, 243, 245, 247, 247, 250, 255, 256, 257, 260, 262, 264, 268, 271, 274, 277, 282, 288, 291.

a) Joy and Joanne arranged these weights in groups 100–120, 120–140, ... 280–300, putting weights of 120, 140, grams, etc., into the lower group. Draw up a table of cumulative frequencies and state the plotting points they should have used. Draw the cumulative frequency graph starting the weights scale at 100 g, and estimate the median, quartiles and interquartile range.
b) Charles and James arranged the weights differently. Their tables showed weights 120, 140, ... 300 g. The entry 160 meant 'weights of just over 150 up to and including 170 g.' Draw up a table using their classification and find the cumulative frequencies

and plotting points. Draw the graph of cumulative frequency on the same axes as you used for *a*). Find the median, quartiles and interquartile range.

c) Comparing the two results, what do you notice? Which gives a truer picture?

*10 In an aircraft factory, the blueprint for a fuselage sub-assembly called for a large number of distance pieces of length 25 ± 0·20 mm. The inspection department examined 100 of these distance pieces drawn at random from the stores, and measured their lengths with a micrometer screw gauge.

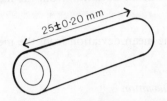

25±0·20 mm

The table shows the lengths they found, all of which were within the required tolerances.

Length (mm)	24·92	24·94	24·96	24·98	25·00	25·02	25·04	25·06	25·08
Frequency	2	8	14	18	23	21	10	3	1

Using these results, copy and complete the table that follows. Draw a cumulative frequency graph showing excess lengths over 24·90 mm on the 'length' axis, and find the median length of the sample of distance pieces and the interquartile range. State the meaning of this latter in terms of the sample of distance pieces examined.

Excess of length over 24·90 mm	Frequency	Cumulative frequency	Plotting point
0·02	2	2	0·03
0·04	8	10	0·05
....			

10E Measures of Dispersion

1 Find the mean of the following sets of numbers:

a) 1, 2, 5, 6, 8, 9, 12, 21   *b*) 6·5, 7, 7, 7·5, 8, 8, 9·5, 10·5

You should find that they both have the same mean but the second group are more closely clustered around the mean than the first group. It is useful to measure the amount of spread. The simplest measure of spread is the range. Write down the range for *a*) and *b*) above.

2 Write down the range for each of the following:

a) 2, 5, 6, 6, 7, 8, 9, 10   *b*) 3, 4, 5, 6, 6, 8, 19, 20   *c*) 4, 5, 5, 6, 7, 8, 34

Why is the range not a very good measure in some cases?

Mean Deviation from the Mean

3 Check that the mean of the numbers 2, 5, 6, 9, 12, 14 is 8. Copy and complete the table. Call all the deviations positive.

Number	Deviation from mean
2	6
5	3
...	...
14	
Total:	

Divide the total deviation by 6 and so find the 'mean deviation from the mean'. What would happen if you did *not* treat all the deviations as positive?

4 Find the mean of the numbers: 6, 5, 7, 8, 9, 10, and use it to find the mean deviation from the mean.

5 Find the mean deviation from the mean of the numbers: 14, 17, 19, 18, 20, 25, 15, 16.

6 Find the mean deviation from the mean of the numbers: 54, 22, 39, 48, 82.

Standard Deviation

7 Check that 19 is the mean of the numbers: 12, 14, 16, 18, 20, 23, 24, 25.
Copy and complete the following table:

Number	Deviation from the mean	(Deviation)^2
12	-7	49
14	-5	25
...	...	...
25		

Add the numbers in the third column and divide the total by 8 (since there are 8 numbers). Find the square root of the result. This is called the standard deviation σ (σ, pronounced 'sigma', is the Greek letter corresponding to our small s).

8 *a)* Using the numbers from question 7, and calling the mean m, find the values of $m-\sigma$ and $m+\sigma$.
b) Write down the numbers in the original list that are:
i) within σ of the mean   *ii)* more than σ from the mean.
c) Work out the values $m-2\sigma$ and $m+2\sigma$. Are any values more than 2σ from the mean?

9 *a)* Find the mean (m) of the numbers: 2, 2, 3, 5, 6, 8, 9, 10, 11, 14.
b) Copy and complete the table:

Number (x)	($x-m$)	($x-m$)^2
2	-5	25
2	-5	25
...	...	...
14		

Add up the numbers in the third column and divide the total by 10. Find the square root of your answer and so write down the standard deviation σ.
c) Work out the values: $m\pm\sigma$ and find the numbers from the list that are more than σ from the mean.
d) Are any values more than 2σ from the mean?

10 *a)* Find the mean and standard deviation of the numbers: 21, 19·5, 18, 19, 20, 19·5.
b) Which of the numbers are within $\pm\sigma$ of the mean?

11 *a)* Find the mean and standard deviation of the numbers: 103, 102, 106, 99, 110.
b) Find which of the numbers are more than $\pm\sigma$ from the mean.

12 *a)* Find the mean and standard deviation of the numbers: 21, 30, 32, 33, 32, 27, 31, 34.
b) Are any of the numbers more than $\pm2\sigma$ from the mean?

13 a) Find the mean and standard deviation of the numbers: 2·8, 2·8, 2·9, 3·1, 2·4, 3·0, 3·2, 3·0, 3·5, 3·3.
b) Which of the numbers are more than ±2σ from the mean?

The Normal Distribution Curve

14

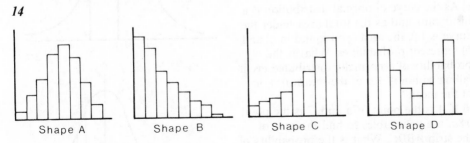

| Shape A | Shape B | Shape C | Shape D |

The diagram shows several typical shapes that can occur in histograms. Most of the histograms studied in 10B approximated to Shape *A*. Shape *B* could represent the errors made by a class of children measuring the weight or height of a given object and calling all errors positive (most of the class would make only a small error). Can you think of distributions giving shapes like *C* and *D*?

If the middle points of the top lines of the rectangles in a frequency diagram or histogram are joined with straight lines, the resulting polygon is called a frequency polygon. For a continuous distribution as the number of strips is increased and their width decreased, the polygon approximates to a curve.
The diagrams below show typical frequency distribution curves corresponding to the histograms above.

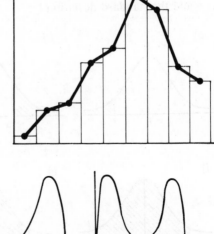

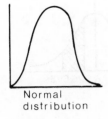
Normal
distribution

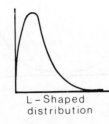
L–Shaped
distribution

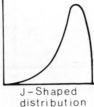
J–Shaped
distribution

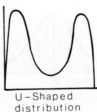
U–Shaped
distribution

The normal distribution is by far the most common and this is the only one that will be studied in this book.

15   The equation of the curve of normal distribution can be calculated but the mathematics are outside the scope of this book. It contains the standard deviation, σ, so the x axis is usually marked in terms of σ, the y axis is taken as the mid ordinate, and the scale is chosen to make the area under the curve 1·0. Tables give the areas between the y axis and selected ordinates.

a) Use your tables to find the area between the y axis and the ordinate AB. What percentage of the distribution lies between $\pm\sigma$?

b) What is the area between OY and CD? What percentage of the distribution lies between $\pm 3\sigma$?

c) As the curve of normal distribution is a histogram, and as the total area under the curve is $1\cdot 0$, the areas you found in *a)* and *b)* represent probabilities. What is the probability of a normally distributed event falling within *i)* $\pm\sigma$ of the mean *ii)* $\pm 3\sigma$ of the mean?

d) If A is the point $0\cdot 7\sigma$ and C is the point $0\cdot 8\sigma$, use your tables to find the area of the strip $ABDC$. What is the probability of a normally distributed event falling between $m+0\cdot 7\sigma$ and $m+0\cdot 8\sigma$, where m is the mean?

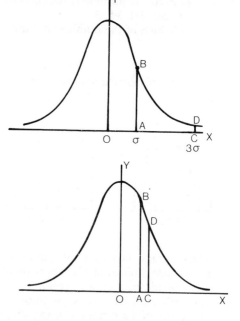

16 Use your tables of areas under the normal distribution curve to find the probabilities represented by the shaded areas in the following. In each case, the mean is 6 and the standard deviation $0\cdot 5$.

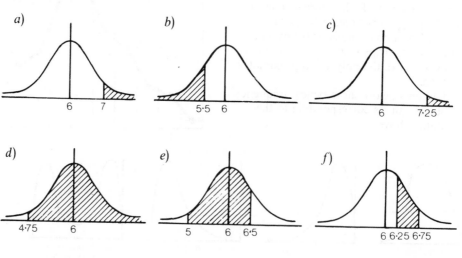

Example From the tables the area shaded vertically is $0\cdot 4772$, so the area shaded horizontally is $(0\cdot 5-0\cdot 4772)=0\cdot 0228$, and the probability is $0\cdot 0228$ or $2\cdot 28\%$.

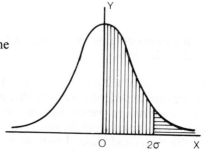

17 The circumferences of young fir trees in a plantation were measured 30 cm from the ground. The mean circumference was 60 cm with a standard deviation of 4 cm. Assuming that the lengths of the circumference were normally distributed about the mean,

a) find the probability of a tree having a circumference of *i*) less than 54 cm
ii) more than 64 cm   *iii*) between 58 and 62 cm   *iv*) between 66 and 70 cm.
b) If there were 3000 trees in the plantation, how many could be expected to have a circumference of *i*) 58 cm or less   *ii*) between 56 and 64 cm?
Give your answers to the nearest 10 trees.

18 Over a period of years the average daily rainfall for the whole of Britain during February was 1·8 millimetres. If the daily rainfall was normally distributed about the mean with a standard deviation of 0·5 mm,

a) on a given day in February what would
be the probability of *i*) less than
1·0 mm   *ii*) more than 2·2 mm
iii) between 0·9 and 1·2 mm?
b) In a non-leap year, on how many days
in February would you expect less than
0·8 mm of rain?
c) Remembering that the information
given concerns rainfall over the whole of
Britain, is your last answer sensible?

19 Nos. 17 and 18 applied the curve of normal distribution to a continuous distribution. Here is an example where it is applied to a discrete distribution.
The number of drawing pins in each of nine boxes is: 41, 37, 40, 39, 42, 38, 43, 42, 38.
Check that the mean and standard deviation are 40 and 2 respectively.
Assume that the number of drawing pins per box is normally distributed about the mean, and that the values you calculate can be taken as the mean and standard deviation of all boxes produced.

a) Find the probability of a box containing less than 36 drawing pins (to apply the normal distribution curve you must think of the distribution being continuous and the number 36 expressing all values between 35·5 and 36·5, i.e. find the area to the left of 35·5).
b) Find the probability of a box containing more than 42 drawing pins (i.e. find the area to the right of 42·5).
c) Find the probability of a box containing *i*) less than 41 pins   *ii*) exactly 41 pins
iii) over 41 pins.
d) Find the value to which the mean must be adjusted so that 4% of the boxes will contain less than 36 drawing pins, assuming that the standard deviation remains unaltered.

20 The numbers of matches in a sample selection of seven boxes are: 52, 53, 54, 51, 55, 50, 56.

a) Find the mean and standard deviation.
b) If the number of matches per box is normally distributed about the mean and the values calculated can be taken as the mean and standard deviation of all boxes produced in the factory from which this sample was taken, find the probability of a box containing less than 50 matches. (Find the area to the left of 49·5.)

c) Printed on each box is the statement 'average contents 52'. What is the probability that a box contains at least this average number?
d) If a box cannot safely hold more than 57 matches, find the approximate number of over-full boxes produced in a batch of 1000. (These would be rejected.)

21 Missiles from a distant launching site are aimed to land on the centre of a city whose diameter is approximately 20 km. If the 'error' is taken as the distance a missile lands from the centre of the city, and this is normally distributed about the mean, zero, with a standard deviation of 5 km,

a) what percentage of the missiles can be expected to land on the city?
b) in what area of the city will the most accurate 50% of the missiles land?

22 The speeds of drivers in the fast lane of a motorway are distributed normally about the mean of 112 km/h, with a standard deviation of 5 km/h. There is a speed limit of 120 km/h.

a) What percentage of the motorists are probably driving between 106 and 114 km/h?
b) What percentage are probably exceeding the speed limit?

23 The number of bathers taking an early morning dip in the Serpentine in Hyde Park, London, during the months of December and January is distributed normally about a mean of 30 with a standard deviation of 3.

a) On a given morning, what is the probability that there will be 28 bathers or less? (Consider 28·5 bathers or less.)
b) If I visited the Serpentine early one morning during those months, what is the probability that I would find *i*) exactly 32 bathers   *ii*) less than 26 bathers?

Note   The normal distribution curve applies strictly to large distributions which are continuous. When applied to small distributions and/or discrete distributions, its results are useful but less accurate.

10F   Miscellaneous

1   The table shows the ages of 100 cars in a car park, judged by the letters on their registration plates:

Age (years)	<1	1–2	2–3	3–4	4–5	5–6	6–7	7+
Number	6	10	17	19	14	9	4	21

Cars aged 2, 3, etc., are included in the higher group.
a) Draw a cumulative frequency graph. Use your graph to estimate, to the nearest half year, the median and the interquartile range of the distribution.
b) Explain why it is not possible to calculate the mean age of the cars.

2   A typing class of 30 were given two short tests and the numbers of mistakes made were tabulated as follows:

No. of mistakes	0–5	6–10	11–15	16–20	21–25	26–30	31–35
No. of pupils (1st test)	2	4	4	5	7	5	3
No. of pupils (2nd test)	1	2	1	8	9	8	1

a) Draw a histogram for each set of results.
b) Estimate the mean for each set of results.
c) Which test did the pupils find most difficult?
d) One student made 20 mistakes in each test. How would you rate her performance?

3   The marks gained by 250 candidates in an examination are shown in the following table:

Marks	10–19	20–29	30–39	40–49	50–59	60–69	70–79	80–89
Frequency	3	7	18	38	52	60	47	25

a) Calculate the mean of the grouped frequencies correct to 1 decimal place.
b) Make a cumulative frequency table for marks of 19 or below, 29 or below, etc.
c) Draw a cumulative frequency graph and use it to find: i) the median mark
ii) the number of candidates who failed if 45 was the pass mark   iii) the lowest mark for distinction if 8% of the candidates were awarded a distinction.

4   A level crossing was closed to traffic five times each day. The number of vehicles held up at each closure was counted in a survey lasting 10 days. The results were as follows:

Number of cars	2	3	4	5	6	7	8	9	10
Times held up	4	3	8	14	6	4	9	0	2

Calculate the mean number of cars held up at each closure and the standard deviation of the distribution.

5   The 'life-times' of 50 small light bulbs used in fairy lights were tested. The results, each to the nearest tenth of an hour, were as follows:

```
70·3  67·6  66·2  70·9  67·3  68·4  70·4  69·3  72·3  70·1
74·7  72·7  70·8  68·7  71·2  70·7  70·6  68·8  69·6  71·0
69·0  69·5  71·4  69·3  71·6  68·4  70·6  66·9  69·6  70·9
71·8  70·5  70·2  67·8  72·7  70·5  69·5  66·6  68·0  67·6
72·5  68·4  69·3  66·2  70·1  69·7  72·6  70·3  69·2  72·6
```

a) Group the results in classes of 1 hour starting with 66·0–66·9 inclusive and plot a histogram to show the frequency distribution obtained.
b) Calculate the mean of the grouped distribution treating each class as located at its mid-interval point.
c) Work out the cumulative frequency and draw a cumulative frequency graph. Use this to estimate the median of the distribution and also the semi-interquartile range.

6   The table shows the distribution of 4000 examination marks.

Mark	0–10	11–20	21–30	31–40	41–50	51–60	61–70	71–80	81–90	91–100
Frequency	50	200	400	700	800	900	600	200	100	50

a) Find the median mark M and the two quartiles Q_1 and Q_3.
b) The examiners decided that 25% of the candidates should be awarded a mark of 45 or less, the median mark should be 55 and the upper quartile 70. Draw a graph and mark the points $(Q_1, 45)$, $(M, 55)$ and $(Q_3, 70)$. Join up with straight lines. Your graph should look something like this. From your graph read off the processed marks corresponding to the original marks of 30, 40, 60 and 80.
c) What is the original mark corresponding to i) the processed mark of 40 which is the pass mark   ii) the processed mark of 80 which is the distinction mark?

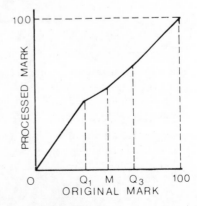

d) Using the results of *c*) and your curve of cumulative frequencies, how many candidates failed and how many got distinction?

7   300 people turn out on a sponsored walk. The walk is a 10 km circuit through woods and fields, with checkpoints every 2 km or so. Walkers may complete the circuit as many times as they like, but must be off the course after twelve hours. At the end of the day the mean distance walked turns out to be 56 km. Assuming the distances walked are normally distributed about the mean with a standard deviation of 10 km, estimate

   a) how many walkers completed eight circuits or more
   b) how many completed 4 circuits or less
   c) how many walked between 64 and 66 km
   d) the value of *d* if half the walkers walked a distance between 44 km and *d* km.

8   The life of an electric light bulb is stated by the makers to be 1000 hours.

   a) If the mean life is 1120 hours and the actual life is normally distributed about the mean with standard deviation 50 hours, state what percentage of bulbs will have a life of *i*) over 1000 hours   *ii*) under 1000 hours.
   b) If the makers wanted to be fairly certain that 99·8% of their bulbs would have the expected life, what figure should they give for the expected life?

9   A class were asked to work out the following sum using slide rules: (22·3 × 29·3)/17·1. If the thirty answers were normally distributed about a mean of 38·2 (which is the correct answer to 3 s.f.) with a standard deviation of 0·5, how many pupils got an answer lying between 37 and 38? How many got more than 39? Give your answers to the nearest whole number.

10   100 pupils in a practical examination were asked to use a potentiometer to measure the resistance of a coil. Although they did not know it, the coils were standard coils with a resistance of 30±0·01 ohms. If their results were normally distributed about a mean of 30 ohms with a standard deviation of 0·4 ohms, how many pupils were within 0·2 ohms of the nominal value of 30 ohms? How many were more than 0·8 ohms out?

11 Similar Triangles

11A Similar Triangles

1 The diagram shows triangle *ABC* with *XY* drawn parallel to *BC*.

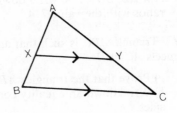

a) Name two pairs of equal angles.
b) What do you know about triangles *ABC* and *AXY*?
c) Write down three equal ratios.
d) If $XY = 2$ cm, $AY = 2.6$ cm, $AX = 1.8$ cm and $BC = 3$ cm, write down the lengths of
i) *AB*   ii) *AC*.

2 Calculate the lengths *a, b, ... h* in the following diagrams.

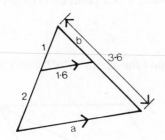

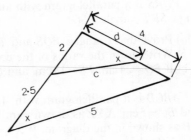

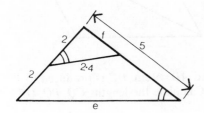

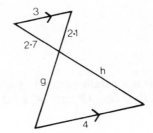

3 *ABCD* is a quadrilateral with diagonal *AC*. Triangles *ACB* and *DAC* are similar, the equal angles being denoted by the order of letters used in naming the triangles.

a) Draw a diagram and mark in the equal angles.
b) Write down three ratios concerning corresponding sides. Are these ratios all equal?
c) If $AC = 6$ cm and $AD = 4$ cm, calculate the length of *BC*.
d) What kind of quadrilateral is *ABCD*?

4 The diagram shows two similar triangles *PQR* and *QSR*. The equal angles are denoted by the order of the letters. Copy the diagram and mark in the equal pairs of angles. Given that $QR = 12$ cm, $PR = 16$ cm and $PQ = 10$ cm, calculate the lengths of the other two sides.

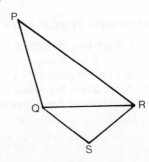

5 The diagram shows triangle *ABC* with angle $B=90°$. The altitude *BD* meets *AC* at *D*. Prove that

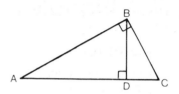

  a) triangles *ABC* and *ADB* are similar
  b) triangles *ADB* and *BDC* are similar.
  c) If $AB:BD=7:4$, write down two more ratios with the value $7:4$.

6 Triangle *ABC* is such that angle *B* is twice angle *A*. *BD*, the bisector of angle *B*, meets *AC* at *D*.

  a) Prove that the triangles *ABC* and *BDC* are similar.
  b) Write down an equality concerning the ratios of the three pairs of corresponding sides.
  c) If $BC=2·4$ cm and $DC=1·8$ cm, calculate the length of *AC*. Hence find the length of *BD*.

7 *PQRS* is a parallelogram with angles *P* and *R* obtuse. *T* is a point on *QS* such that angle $SRT=$ angle *PSQ*.

  a) Prove that triangles *PQS* and *TSR* are similar.
  b) Write down the ratios of the corresponding sides.
  c) If $PQ=24$ cm, $PS=18$ cm, and $QS=36$ cm, calculate the lengths of *ST* and *TR*.

8 *ABCD* is a parallelogram with $AB=6$ cm and $BC=5$ cm. *XYZ* is drawn parallel to *AB* as shown in the diagram. If $XY=3·6$ cm and $YD=4·5$ cm, calculate the lengths of *BY* and *BZ*.

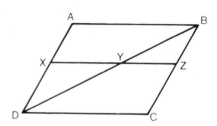

9 *ABCD* is a parallelogram in which $AB=5$ cm and $AD=9$ cm. *DC* is produced to *P* so that $CP=1$ cm. *AP* cuts *BC* at *Q* and $AD=AQ$. Calculate the lengths *CQ*, *BQ*, *QP* and *AP*.

10 *LMN* is a triangle with *XY* drawn parallel to *NM* and *XZ* parallel to *LM* as shown in the diagram.
If $NM=8$ cm, $XY=3$ cm, $XZ=4$ cm and $LN=7·2$ cm, calculate the lengths *LY*, *LX* and *XN*.

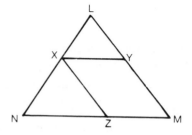

11B The Rectangle Property of a Circle

1 *AB* and *CD* are two chords of a circle which intersect at *P* as shown in the diagram. Prove that triangles *ACP* and *DBP* are similar. Write down the ratios of the corresponding sides and hence prove that $AP.PB=CP.PD$.

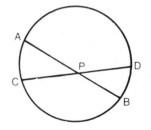

2 If the two chords AB and CD of a circle are such that they do not intersect inside the circle but are both produced and meet at Q, outside the circle, prove that the triangles ADQ and CBQ are similar. By writing down the ratios of corresponding sides prove that $AQ \cdot QB = CQ \cdot QD$.

3 Calculate the lengths $a, b, c, \ldots i$ in the following diagrams:

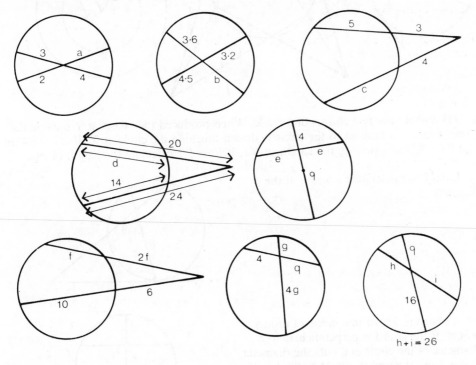

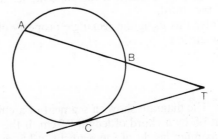

4 A tangent touches a circle at C and meets chord AB produced at T. Prove that triangles ATC and CTB are similar. By writing down the ratios of the corresponding sides, prove that $CT^2 = AT \cdot TB$.

5 Calculate the lengths $a, b, \ldots k$ in the following diagrams:

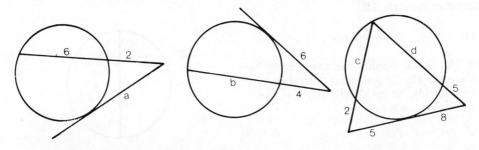

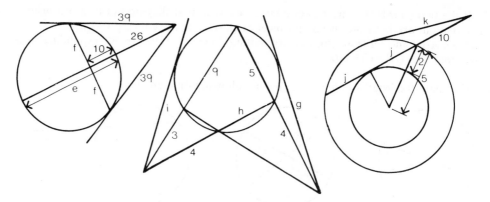

6 *PQ* and *RS* are two chords of a circle. When produced they meet at *T* outside the circle. From *T* a tangent to the circle is drawn touching the circle at *X*. If $XT = 8\cdot4$ cm and $ST = 4\cdot2$ cm, calculate the length of *RT*. If $RS = PT$ calculate the length of *PQ*.

7 Calculate the lengths *a* and *b* in the diagram.

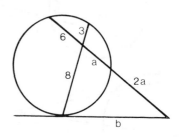

8 A chord is drawn in a circle of radius 6·5 cm. If the chord is perpendicular to a diameter of the circle and cuts the diameter 2·5 cm from the centre, calculate the length of the chord.
(Solve this using the rectangle property of the circle.)

9 The diagram shows a segment of a circle cut off by a chord of length 9·0 cm. If the maximum height of the segment is 2·5 cm, find the radius of the circle of which this forms a part.
(*Hint* Complete the circle and draw in the diameter through *AB*.)

10 A circular window has two glazing bars as shown. The longer bar is central and is perpendicular to the shorter bar. Find the radius of the window.

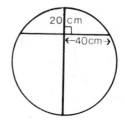

＊ 11 A furniture removals van 2·4 m wide and 2·9 m high wishes to go under a bridge which is an arc of a circle. The width of the arch at road level is 8 m and the height of the centre of the arch above the road is 3·2 m. Can the van get through?

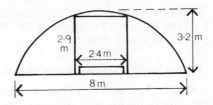

12 a) *ABCD* is a rectangle. *AB* is produced to *E* so that *BE* = *BC*. A circle is drawn on *AE* as diameter. If *AB* = 5 cm and *BC* = 3 cm, find *BF*.

b) Draw accurately a rectangle 6 cm by 4·5 cm and use the above construction to find by actual drawing the side of a square equal in area to the given rectangle.

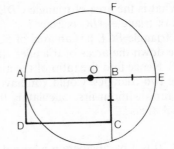

13 *ABC* is an isosceles triangle and *AB* = *AC*. *BC* is a tangent to a circle at *C*, and this circle passes through *A* and cuts *AB* at *L*. The line *LMN* is parallel to *BC* as shown in the diagram. If *BC* = 18 cm and *BL* = 12 cm, calculate the lengths of *AL*, *LM* and *MN*.

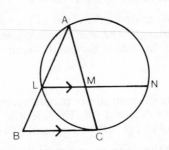

14 *ABC* is a triangle inscribed in a circle. The bisector of angle *C* cuts *AB* at *D* and meets the circle at *E*. Prove that triangles *EBD* and *ECB* are similar, and show that $EB^2 = CE \cdot DE$. If *EB* = 4·2 cm, *CE* = 6·3 cm and *DB* = 4·9 cm find the lengths of *DE* and *AD*.

15 Two circles touch externally at *A*. The tangent at *T* on one circle cuts the other at *B* and *C* so that *T*, *B*, *C* lie in that order. *CA* produced cuts the first circle at *R*. The common tangent at *A* meets *TB* at *S*.

a) Prove i) angle *ACB* = angle *BAS*   ii) angle *ART* = angle *TAS*
iii) angle *TAR* = angle *TAB*.

b) Hence prove that triangles *TAB* and *RAT* are similar and show that $AT^2 = AR \cdot AB$.

11C More Ratios and Similarities

Areas of Similar Triangles

1 The diagram shows triangle *ABC* with *XY* drawn parallel to *BC*. If *AY* = 1·3 cm, *AC* = 5·2 cm and *BC* = 4 cm, find the length of *XY*.
If the area of triangle *AXY* is 0·6 cm², calculate the area of triangle *ABC*.

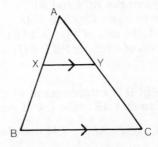

2 Using the same diagram as in question 1, if $AY = 3.2$ cm, $YC = 1.6$ cm and $XY = 2$ cm, calculate the length of BC. If the area of triangle ABC is 12 cm², calculate the areas AXY and $XYCB$.

3 In the diagram AB is parallel to DE.

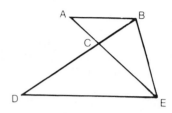

a) If $AB = 2$ cm and $DE = 5$ cm, write down the ratio of the areas of triangles ABC and CDE.
b) What is the area of triangle CDE if the area of triangle ABC is 2 cm²?
c) If triangle BCE has an area of 5 cm², write down the areas of triangles ABE and BDE. Hence find the ratio of the areas of these two triangles. Could you have found this ratio without first calculating the actual areas?

4 $ABCD$ is a parallelogram whose diagonals intersect at E. A line is drawn from E parallel to DA to meet AB at F.

a) Write down the ratios of the areas of i) triangle AFE and triangle ABC
ii) triangle AFE and parallelogram $ABCD$.
b) If the area of the parallelogram $ABCD$ is 30 cm², what is the area of triangle AFE?

5 The diagram shows parallelogram $PQRS$ with the diagonal PR. The line ABC is parallel to PQ, $AP = 1.7$ cm and $SP = 5.1$ cm.

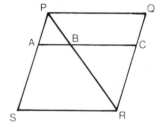

a) Write down the ratios of the areas of
i) APB and BCR ii) APB and PSR
iii) BCR and PSR iv) BCR and $ABRS$.
b) If the area of $PQRS = 54$ cm², what is the area of $ABRS$?

6 The diagram shows triangle PRT with QS drawn parallel to RT.

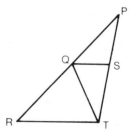

a) If $QS = 2.7$ cm and $RT = 7.2$ cm, write down the ratios of the lengths of
i) PQ and PR ii) RQ and RP
iii) PS and ST.
b) Hence write down the ratios of the areas of i) triangles PQS and PRT
ii) triangles RQT and RPT
iii) triangles PSQ and TSQ.
c) If the area of PQS is 4.5 cm², find the areas of triangles PRT, RQT and TSQ.

7 $ABCD$ is a trapezium in which AB is parallel to DC. The diagonals cut each other at E so that $AE:AC = 1:4$. If the area of triangle DEC is 63 cm², what is the area of

a) triangle AEB b) triangle AED c) triangle ABC?

8 *a*) From the preceding examples we can now say that the areas of similar triangles are proportional to the squares on corresponding sides. Are there any exceptions to this rule?

b) In the figure the irregular hexagon *ABCDEF* has been enlarged by a factor of 1·5 to give *PQRSTU*.

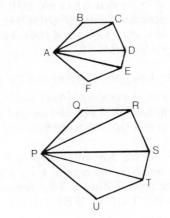

i) Are triangles *ABC* and *PQR* similar?

ii) Is the area of triangle *PQR* $(1·5)^2$ times the area of triangle *ABC*?

iii) Are your answers to i) and ii) true for the pairs of triangles *ACD* and *PRS*, *ADE* and *PST*, *AEF* and *PTU*?

iv) Is it true to say that the area of the whole hexagon *PQRSTU* is $(1·5)^2$ times the area of the hexagon *ABCDEF*?

v) Is it true to say that the areas of similar polygons are proportional to the squares on corresponding sides?

The 'Bisector Theorems'

9 The diagram shows triangle *ABC* with a point *D* in *AB* such that *CD=DB*. The line through *D* parallel to *BC* meets *AC* at *E*. Prove that

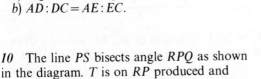

a) angle *ADE*=angle *EDC*

b) *AD:DC=AE:EC*.

10 The line *PS* bisects angle *RPQ* as shown in the diagram. *T* is on *RP* produced and *TQ* is parallel to *PS*. Prove that

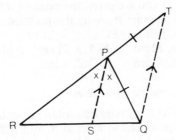

a) triangle *PQT* is isosceles

b) *RP:PQ=RS:SQ*.

If *PR*=5·4 cm and *PQ*=4·5 cm, find the ratio *RS:SQ*.

11 Using the diagram and the result of question 10, find the length of

a) *PR* given that *RS:SQ*=5:3 and *PQ*=3·6 cm.

b) *PR* given that *RS:RQ*=4:7 and *PQ*=4·2 cm.

c) *PQ* given that *PR*=4·8 cm and *RS:RQ*=3:10.

12 The diagram shows triangle *ABC* with *AB* produced to *F*. *BE=BD* and *ED* is parallel to *BC*. Prove that

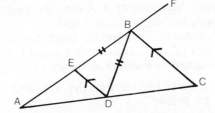

a) angle *DBC*=angle *CBF*

b) *AC:CD=AB:BD*.

13 The line QA bisects the exterior angle RQS of triangle PQR. The line BR is parallel to QA.

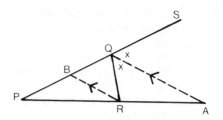

a) Prove that i) triangle BQR is isosceles   ii) $PQ:QR=PA:AR$.
b) If $PQ=8{\cdot}4$ cm and $QR=3$ cm, find the value of $PA:AR$.

14 Using the diagram and the results of question 13,

a) find PQ if $QR=3$ cm and $PA:AR=5:2$.
b) find QR if $PQ=7{\cdot}2$ cm and $PA:AR=8:3$.
c) find QR if $PQ=10{\cdot}8$ cm and $PR:RA=11:7$.

15 In the diagram, the point X divides AB *internally* in the ratio $2:1$, i.e. $AX:XB=2:1$. The point Y divides AB *externally* in the ratio $2:1$, i.e. $AY:YB=2:1$.

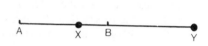

a) If $AB=12$ cm, state the lengths of i) AX, ii) XB, iii) AY, iv) YB.
b) If $AY:YB=2:3$, i) draw a rough diagram showing the position of Y
ii) state the lengths of AY and YB if $AB=15$ cm.

16 From the preceding examples we can now see that in a triangle ABC, if $AB:AC=m:n$ then AD, the internal bisector of angle A, divides BC internally in the ratio $BD:DC=m:n$; and AE, the external bisector of angle A, divides BC externally in the ratio $BE:EC=m:n$.

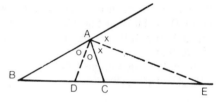

In a) to f) you are given the ratio of $AB:AC$ and the length of BC. In each case draw a rough diagram showing the position of AD and AE, and give the lengths of i) BD, ii) DC, iii) BE, iv) EC.

a) $6:5$, 22 cm   b) $4:3$, 21 cm   c) $3:4$, 21 cm   d) $2:5$, $17{\cdot}5$ cm
e) $1:4$, 15 cm   f) $2:2$, 8 cm

Similar Triangles in Circles

17 $ABCD$ is a cyclic quadrilateral with AD and BC produced to meet at X and AB and DC produced to meet at Y as shown.

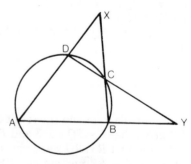

a) Name two pairs of similar triangles in the diagram.
b) If $XD:XB=5:9$ and $YB:YD=5:11$, find the values of i) $DC:AB$, ii) $CB:DA$.

18 The diagram shows two circles which touch each other at A. A common tangent meets the two circles at E and C respectively. The common tangent at A meets EC at T. The chords EA and CA are produced to B and D as shown.

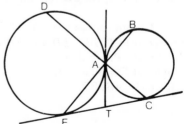

a) Prove that the triangles ABC and ACE are similar.

b) By writing down the ratios of the corresponding sides, prove that $AC^2 = AB . AE$.
c) Prove that triangles ABC and AED are similar.
d) Given that $AB = 2.7$ cm and $AE = 4.8$ cm, calculate the lengths of AC and AD.

＊19 The tangent to the circle at A meets the chord CB produced at T. D is a point on AT such that $AB^2 = AC . BD$. Angles TAC and TDB are obtuse.

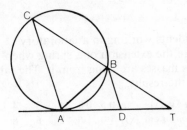

  a) By finding a pair of similar triangles prove that AC and DB are parallel.
  b) Name two more pairs of similar triangles.

20 Two circles touch each other at A as shown in the diagram, and AT is the common tangent at A. The smaller circle passes through Y, the centre of the larger circle. The chord AB of the outer circle cuts the inner circle at X. If BC is a tangent to the inner circle, touching it at Y,

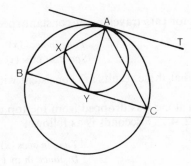

a) prove that AT is parallel to BC.
b) prove that triangles AXY and BAC are similar
c) find the value of $XY : AC$.

12 Variation

12A

1 Students working in a laboratory measure the extension of a spring when various masses are hung from it. The table shows the results.

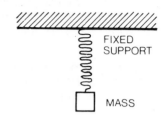

FIXED SUPPORT

MASS

Mass in kg (y)	0·5	0·8	1	3	4	6
Extension in cm (x)	1·0	1·6	2	6	8	12

Show that these results are consistent with a law in the form $y=kx$, and find k.

2 Motor cars travelling at constant speeds were timed over a given distance.

Time in secs (x)	20	30	40	60	80
Speed in m/s (y)	24	16	12	8	6

Show that these results are consistent with a law in the form $y=k/x$, and find k.

3 If a stone was dropped from the top of a tower 180 m high, the distance it would fall in 2, 4,...6 seconds is as follows:

Time in secs (x)	2	4	5	6
Distance in m (y)	20	80	125	180

Find the equation which relates x and y.

4 The time for one complete swing of a simple pendulum 2·25 m long was measured by an accurate timing device. The experiment was repeated for pendulums of other lengths.

Length in m (x)	2·25	1·96	1·69	1·44	1·00
Time in secs (y)	3	2·8	2·6	2·4	2·0

Show that these results are consistent with a law in the form $y=k\sqrt{x}$, and find k.

5 A number of different shaped boxes are used by a manufacturer for packing. All the boxes have square bases and the table shows the length of the side of the square base and the height of each type of box.

Length of base in cm (x)	75	100	120	150	160
Height in cm (y)	256	144	100	80	56·25

Show that all these measurements except one are consistent with a relationship in the form $y=k/x^2$.
Find k and amend the inconsistent value in the table to make it consistent with the formula.
Deduce a simple fact about all the boxes.

6 In question 1, we can either say that y is proportional to x or that $y \propto x$. State the relationship between x and y in these two different ways for each of questions 2, 3, 4, and 5.

7 Plot the information given in question 1 on a small scale graph to show the relationship between x and y. Repeat for questions 2 to 5.

8 In each of the following x and y are related to one another by a law in the form $y \propto x^n$. Find the value of n in each case.

a) y is the area of a circle and x its radius.
b) x is the length of an edge of a cube and y is its mass.
c) x is the height of a column of given width in a histogram and y is its area.
d) x is the volume of water in a rectangular tank and y the depth of the water.
e) x is the depth of water in a rectangular tank and y is the length of time during which the tank (originally empty) has been filling at a constant rate.
f) x is the length of a rectangle of given area and y is the width.
g) x is the volume of a sphere and y is its radius.
h) x is the constant speed of a car and y is the time which it takes to cover a certain fixed distance.
i) x is the time taken from rest to reach a speed y, the acceleration being constant.
j) x is the height of a cylinder with a certain fixed capacity and y is its radius.

12B

1 $y \propto x$ and $y=3$ when $x=1$. Find the law which connects x and y and find the value of x when $y=7 \cdot 5$.

2 y varies directly as x and $y=2$ when $x=3$. Find the law which relates x and y. Calculate y when $x=9$ and calculate x when $y=0 \cdot 5$.

3 $y \propto x^2$ and $y=3$ when $x=2$. Find the relationship between x and y. Calculate the value of y when $x=1$ and the value of x when $y=6$.

4 y varies as the square of x and $y=9$ when $x=6$. Find the law which relates x and y. Calculate the value of y when $x=4$ and calculate the value of x when $y=2$.

5 $y \propto x^3$ and $y=3$ when $x=1$. Find the value of y when $x=2$.

6 $y \propto \sqrt{x}$ and $y=1$ when $x=4$. Find the relationship between x and y and calculate the value of y when $x=6 \cdot 25$.

7 y varies as the square root of x and $y=2$ when $x=2 \cdot 25$. Find the value of y when $x=9$ and the value of x when $y=1$.

8 $y \propto 1/x$ and $y=2$ when $x=2$. Find the law which relates x and y and calculate the value of y when $x=0 \cdot 5$.

9 y varies inversely as x and $y=1$ when $x=6$. Find the relationship between x and y and calculate the values of x when $y=1 \cdot 5$ and when $y=9$.

10 y varies inversely as the square root of x and $y=5$ when $x=4$. Find the equation which relates x and y and calculate the value of y when $x=16$.

11 y and x are connected by an 'inverse square law', i.e. $y \propto 1/x^2$ and $y=3$ when $x=1$. Find the value of y when $x=9$ and the value of x when $y=27$.

12 $y \propto 1/\sqrt{x}$ and $y=2$ when $x=25$. Find the law which relates x and y, and calculate the value of x when $y=4$. Calculate the value of y when $x=10$.

13 $A \propto l$ and $A = 36$ when $l = 9$. Find the law which relates A and l and calculate the value of A when $l = 12$.

14 $V \propto r^2$ and $V = 96$ when $r = 4$. Find r when $V = 150$ and when $V = 121 \cdot 5$.

15 $T \propto \sqrt{l}$ and $T = 6$ when $l = 6 \cdot 25$. Calculate the value of l when $T = 4$, and the value of T when $l = 3 \cdot 75$.

16 l varies inversely as b and $l = 9$ when $b = 8$. Calculate the values of l when $b = 6$ and when $b = 12$.

17 h varies inversely as the square of r and $h = 18$ when $r = 4$. Calculate the value of h when $r = 12$ and the value of r when $h = 8$.

18 $l \propto 1/\sqrt{h}$ and $l = 14$ when $h = 25$. Find the relationship between l and h and calculate the value of l when $h = 12 \cdot 25$ and the value of h when $l = 21$.

19 $V \propto r^3$ and $V = 180$ when $r = 3$. Find the law which relates V and r, and calculate V when $r = 4 \cdot 5$. When $V = 1440$ calculate r.

20 x varies as the cube root of M and $x = 2 \cdot 5$ when $M = 125$. Calculate x when $M = 512$.

21 For each of the following state the change in y if x is doubled:

  a) $y \propto x$ b) $y \propto 1/x$ c) $y \propto x^2$ d) $y \propto x^3$ e) $y \propto 1/x^2$

22 For a body starting from rest and moving with constant acceleration, the distance covered is proportional to the square of the time. If it takes t seconds to travel 10 metres, how long will it take to travel 40 metres?

23 For a body starting from rest and moving with constant acceleration, the velocity attained is proportional to the square root of the distance travelled. If a velocity of 5 m/s is reached in a distance of s metres, what distance is covered in reaching a speed of 20 m/s?

24 For boxes of a certain fixed capacity, the length of the side of the square base is inversely proportional to the square root of the height. State what change takes place

  a) in the height if the base length is doubled
  b) in the base length if the height is doubled.

25 If $y \propto x$ and $x \propto \sqrt{z}$, write down the relationship between y and z.

26 If y varies as the square root of x, and x varies inversely as z^2, write down the connection between y and z.

27 If $y \propto x^2$ and also y is inversely proportional to z, write down the relationship between x and z.

28 If $y \propto 1/\sqrt{x}$ and also $y \propto z^2$, state how x varies with z.

29 If the velocity reached by a moving body is proportional to the time taken and the distance covered is proportional to the square of the time taken, state how the velocity varies with the distance and how the distance varies with the velocity.

30 If $y \propto 1/x^2$ and $z \propto \sqrt{x}$, state how y varies with z.

1 The cost of producing a toy motorcar is partly constant and partly depends on the number produced. This can be expressed as $C = a + bn$ where C is the cost in pence and n the number of cars produced. If $C = 85$ when $n = 100$ and $C = 76$ when $n = 1000$, find the values of a and b. What would be the price of a car if 500 were produced?

2 The cost of producing a model of a road layout is partly constant and partly varies as the square of the number of models produced. If 10 models cost £20 each to produce and 5 models cost £27·50 each, what would be the cost of a model if 8 were produced?

3 The estimated cost of excavating a length of roadway is partly constant and partly varies as the square of the depth excavated. If the estimated cost of excavating is £63 500 for a depth of 30 cm and £74 000 for a depth of 40 cm, what would be the estimated cost of excavating to a depth of 50 cm?

4 The charge for feeding each camper in a boys' camp is partly constant and partly varies inversely as the number of campers. If the daily charge per head is 88 pence when there are 50 campers and 85 pence when there are 80 campers, what would be the number of campers when the daily charge per head was 84 pence?

5 y varies partly as x^2 and partly as $1/x$. This can be written as $y = ax^2 + b/x$ where a and b are constants. If $y = 54$ when $x = 4$, and $y = 24$ when $x = 2$, find y when $x = 3$ and also find x when $y = 112$.

6 z varies jointly as x and y. This can be written $z \propto xy$ or $z = kxy$ where k is a constant. If $z = 48$ when $x = 3$ and $y = 8$, find the value of k. Find also the value of z when $x = 4$ and $y = 6$, and find the value of y when $z = 128$ and $x = 16$.

7 The relation in question 6 can also be written $z_2/z_1 = (x_2/x_1)(y_2/y_1)$ where x_1, y_1 and z_1 are one set of values of x, y and z, and x_2, y_2 and z_2 are another set.

   a) If x is doubled and y is trebled, what happens to z?
   b) If z is halved and y doubled, what happens to x?

8 p varies inversely as the square of q and directly as r. Write this in the two ways given in question 6. Using suffixes 1 and 2, write it also in the way given in question 7. What happens to p when q is doubled and r is halved? What happens to r when q is halved and p is trebled?

9 $y \propto z/\sqrt{x}$. If $y = 36$ when z is 6 and x is 4, find y when z is 10 and x is 9.

10 The gravitational force between two bodies of masses m_1 and m_2 at a distance r apart is proportional to the product of m_1 and m_2, and inversely proportional to r^2. In a certain set of units the force F is 128 when the masses are 20 and 40 and r is 5. State the law connecting F with m_1, m_2 and r. If the force between two bodies of masses 40 and 90 is 100, what is the distance between them?

13 The Earth's Surface

13A   Latitude and Longitude

1 a) The earth's surface is approximately spherical. In what ways does it differ from a true sphere?

b) It is spinning about an axis through the North and South Poles, marked *N* and *S* in the diagram. *O*, the centre of the earth, lies on this axis, and a plane through *O* perpendicular to this axis cuts the earth's surface in a circle known as the equator (*ABCD* in the figure). Where is the centre of this circle?

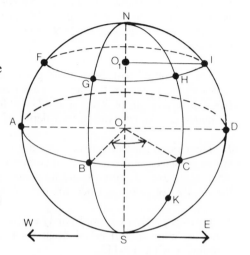

c) The length of a quarter of the circumference of the equator is 10 000 km. What is the whole circumference? Taking π as 3·142, what is the radius of the earth (to the nearest 10 km)?

d) How many degrees are there in a full circle? How many minutes are there? What is the length in kilometres of one minute of arc of the equator?

e) Distances at sea are often measured in nautical miles. A nautical mile is one minute of arc. How many kilometres are there in a nautical mile?

2 a) A circle on the surface of the earth with its centre at *O*, the centre of the earth, is known as a great circle. Name three great circles in the diagram in question 1.

b) Any circle passing through *N* and *S* is known as a meridian of longitude. Name two meridians of longitude in the given figure. Is every meridian of longitude a great circle?

c) A circle drawn in a plane parallel to the equator and lying on the earth's surface is called a parallel of latitude. *i)* Name two parallels of latitude in the given figure. *ii)* What is the smallest possible parallel of latitude? *iii)* What is the largest?

3   The meridian of longitude which passes through Greenwich is known as the meridian of Greenwich. It is marked *NGBS* in the diagram. *NHCS* is another meridian. These meridians cut the equator at *B* and *C* respectively, and the angle *BOC* is called the angle of longitude. It is measured from the meridian of Greenwich East or West.

a) What is the greatest possible angle of longitude? What is the least?

b) What can you say about the meridians *i)* 180° E and 180° W   *ii)* 90° E and 90° W   *iii)* 50° E and 130° W?

4 a) In the diagram in question 1, *H* is a point on the meridian *NHCKS*. Its latitude is defined by the angle *HOC*. *K* is another point on the same meridian. Name the angle of latitude of *K*.

b) What is the locus of all points with the same latitude?

c) If angle *HOC* is 43°, *H* has latitude 43° N. If angle *KOC* is 54°, what is the latitude of *K*?

5 To construct a model to show latitude and longitude, first mark out three equal circles with diameters about 8 or 10 cm, and also a single quadrant, on thin cardboard or stiff paper. Draw rectangles enclosing the lower half of each of the first two circles and cut them out as shown in the diagram.

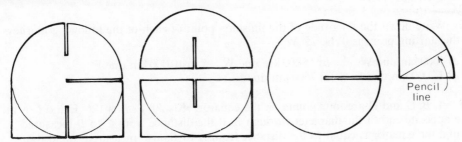

Pencil line

The second circle is cut into two parts as shown (a semi-circle and a rectangle). The long slots are one radius in length and the short slots half a radius.

The model is easily assembled and held together with sellotape along the seams. The quadrant is hinged with sellotape on the side *NO* and can be rotated about the vertical axis *NOS*. The whole model can be mounted on a square of cardboard if desired.

As the quadrant is rotated, the point *R* traces out an arc of a parallel of latitude. The angle of latitude is *ROQ*. The angle of longitude is *AOQ*, where *NAS* is taken as the meridian of Greenwich.

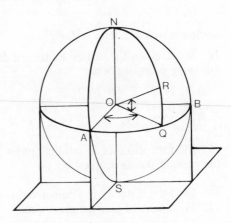

Distances round the Equator

6 You are given the longitudes of pairs of points which lie on the equator. In each case find the angle subtended at the centre of the earth by the arc which joins the two points.

a) 30° E, 50° E   b) 23° E, 77° E   c) 18° W, 156° W
d) 117° E, 117° W   e) 18°47′ E, 136°56′ E   f) 126°12′ W, 44°48′ E
g) 48° E, 168° W   h) 71°11′ E, 45°19′ W   i) 146° E, 172°15′ W
j) 90°20′ E, 89°40′ W

7 Find the distance in nautical miles measured along the equator, between the pairs of points in questions 6a)...e).

8 Find the distance in km measured along the equator between the pairs of points in question 6f)...j). Give your answers correct to 3 s.f.

9 *P* is the point whose longitude is 55° E. Give the longitude of the places which are
a) 30° W of *P*   b) 50° E of *P*   c) 120° E of *P*   d) 95° W of *P*   e) 140° E of *P*.

10 Give the longitude of the finishing point of each of the following journeys along the equator if they all start at the point on the equator whose longitude is 15° W.

a) 2100 nautical miles due E   b) 4320 nautical miles due E
c) 4500 nautical miles due W   d) 9900 nautical miles due W
e) 1350 nautical miles due E

11 *P* and *Q* are at opposite ends of a diameter of the earth. You are given the position of *P*. Write down the position of *Q*.

a) 0° N, 60° E   b) 0° N, 12° W   c) 0° N, 115° W   d) 0° N, 175° W
e) 0° N, 110° 45′ E

12 Write down the longitude of the finishing point of each of the following journeys if the starting point is 0° N, 75° W.

a) 5000 km due W   b) 15 000 km due W   c) 10 000 km due E
d) 14 000 km due E   e) 7500 km due E

13 *A*, *B*, *C* and *D* are four points on the equator. *A* is 30° E, *B* is 125° E and *C* is at the opposite end of the diameter through *B*. If the distance from *C* to *D* measured round the equator is equal to the distance from *A* to *B*, also measured round the equator and in the same direction, write down the longitude of *D*.

Distances round Circles of Longitude

14 Write down the angles subtended at the centre of the earth by the arcs joining the following pairs of points:

a) 36° N, 27° E; 56° N, 27° E   b) 15° N, 84° E; 73° N, 84° E
c) 38° N, 45° E; 17° 11′ N, 45° E   d) 27° 53′ N, 18° W; 32° 7′ S, 18° W
e) 65° 12′ N, 100° W; 47° 18′ S, 100° W

15 Find the distance in nautical miles, measured along a circle of longitude, between the pairs of points in question 14.

16 Find the distance in kilometres between the points 10° S, 115° W and 50° N, 115° W, measured along a great circle.

17 The great circle distance between *A* (0° N, 25° E) and *B* (0° N, 67° W) is the same as the distance along the meridian between *P* and *Q*. If *P* is 45° N, 70° E, give two possible positions of *Q*.

18 What is the final position of a plane which flies 6500 km due N starting at the point 15° S, 28° W?

19 A plane flies from *A* (57° N, 60° E) along a great circle, over the North Pole to reach *B* (57° N, 120° W). How long is its journey measured in nautical miles?

Distances round Circles of Latitude

20 The diagram shows the plane of the meridian *NPASB*. *O* is the centre of the earth. O_1 is the centre of the circle of latitude on which *P* lies.

a) If the position of *P* is 50° N 30° E, name two angles of 50° in the diagram.
b) If *OA* is of length *R*, how long is *OP*?
c) How long is O_1P?
d) What is the circumference of the circle of latitude? Give your answer in terms of *R* and *π*.

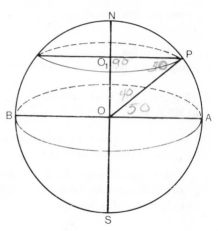

21 For the great circle of the earth, radius R, one minute of arc is one nautical mile. What will be the length of one minute of arc for a circle of latitude $x°$ N?

22 Find the distance between each of the following pairs of points measured along the parallel of latitude on which they lie. Give your answers in nautical miles correct to 3 significant figures. In each case select the shorter of the two possible distances.

a) 27° N, 13° E; 27° N, 53° E
b) 53° N, 27° E; 53° N, 76° W
c) 15° S, 156° E; 15° S, 120° W
d) 25° S, 170° W; 25° S, 10° W
e) 64° N, 35° 30′ W; 64° N, 158° E

23 Find the distance between the following pairs of points measured along the circle of latitude on which they lie. Give your answers in kilometres correct to 3 significant figures. In each case find the shorter of the two possible distances.

a) 15° N, 10° E; 15° N, 170° W
b) 35° N, 80° W; 35° N, 140° W
c) 40° S, 37° 30′ E; 40° S, 30° W

24 *a)* A ship fitted with sonar equipment for measuring the depth of the ocean bed sails on a special survey 3000 km westward along a parallel of latitude starting from 48° N, 10° W. Give the position of its finishing point.
  b) Repeat *a)* for *i)* a ship sailing east for 13 000 km starting from 60° S, 120° W *ii)* a ship sailing west for 9500 km starting from 20° N, 120° W. Ignore any detours due to small islands, etc., and give your answers to the nearest half degree.

Note In numbers 25 to 28 you are given the radius of the earth in kilometres. Work 'ab initio' from this without using the relationship between a minute of arc and a nautical mile.

25 Taking the radius of the earth as 6370 km, find the shorter distance between the following pairs of points, measured along the great circle on which they lie:

a) 0° N, 63° E; 0° N, 58° W
b) 0° N, 63° E; 73° N, 63° E
c) 0° N, 63° E; 46° N, 117° W
d) 23° S, 15° W; 46·5° N, 15° W
e) 68° N, 45° W; 68° N, 135° E

Give your answers in kilometres, correct to 3 s.f.

Example The distance in *a)* is $2\pi \times 6370 \times \left(\dfrac{63+58}{360}\right)$.

26 Taking the radius of the earth as 6370 km, find the shorter distance between the following pairs of points measured along the parallel of latitude on which they lie:

a) 28° N, 46° W; 28° N, 19° W
b) 16° N, 11° W; 16° N, 82° E
c) 59° S, 49° E; 59° S, 49° W
d) 73° N, 81° W; 73° N, 105° E
e) 34° S, 34° E; 34° S, 170° W

Give your answers in kilometres, correct to 3 s.f.

Example The distance in *a)* is $2\pi \times 6370 \times \left(\dfrac{46-19}{360}\right)\cos 28°$.

27 Taking the radius of the earth as 6370 km, find the distance between each of the following pairs of points measured along the meridian of longitude or the parallel of latitude on which they lie:

a) 18° N, 90° W; 18° S, 90° E
b) 36° N, 150° E; 80° S, 150° E
c) 65° S, 118° W; 65° S, 49° E
d) 52° N, 23° W; 52° N, 150° E
e) 27° N, 36° W; 27° S, 144° E

28 *a)* A submarine contacts a supply ship at 50° N 20° W and sails 1000 km due south. What is its new position?

b) If instead it had sailed 800 km due west, what would its position have been then?

Great Circle Sailing

*29 The shortest distance between two points on a plane is the straight line joining them. The shortest distance between two points on the surface of a sphere is the arc of the great circle through the points. Ships sailing on the open ocean usually follow a great circle course.

The calculation of 'great circle distances' between two points is in general beyond the scope of this book, but two particular cases are reasonably simple, i.e. when the two points lie on a meridian of longitude or on the equator, and when the two points lie on a parallel of latitude. The first case has already been studied.

For the second case, proceed as follows:

The figure shows two points A and B with longitudes 30° E and 80° E, both on the same parallel 40° N.

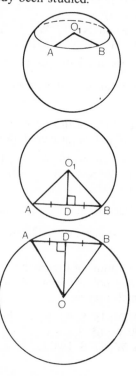

a) Given that OA, the radius of the earth, is 6370 km, find O_1A, the radius of the parallel of latitude.

b) The second figure is in the plane of the parallel of latitude through A and B. What are the angles AO_1B and AO_1D?

c) What is the length AD?

d) The third figure is in the plane of the great circle through A and B. Knowing the lengths of AD and AO, calculate angle AOD and angle AOB.

e) What is the length of the great circle arc AB?

f) Using one of the methods already used in this chapter, calculate the distance between A and B along the parallel of latitude.

g) Which is the shorter, the distance in *e* or the distance in *f*?

*30 Repeat question 29 for the following pairs of points, giving as your answers the distances in km along the great circle and along the parallel of latitude:

a) 20° N, 37° W; 20° N, 121° W   b) 45° N, 37° W; 45° N, 121° W
c) 70° N, 37° W; 70° N, 121° W   d) 32° N, 18° W; 32° N, 40° E
e) 56° S, 11° W; 56° S, 98° E

Comparing your answers to a, b and c, do the great circle distance and the distance along the parallel of latitude differ more widely as you move away from the equator?

13B   Miscellaneous

Note   Take the radius of the earth as 6370 km and give your answers correct to 3 s.f. where appropriate. All latitudes and longitudes quoted are approximate only.

1   Find the distance in kilometres between Amoy (24° N, 118° E) and Abu Dhabi (24° N, 54° E) measured along the circle of latitude on which they both lie.

2   Travelling due south from Amoy for 5000 km one reaches Pilbara, Australia. Give the position of Pilbara.

3   Find the distance between Pike's Peak (39° N, 105° W) and Amritsar (31° N, 75° E) measured along a line of longitude via the North Pole.

4   Stanleyville and the island of Andros in the Aegean Sea are both in longitude 25° E. Stanleyville is on the equator and the latitude of Andros island is 38° N. Find the longitude of a point P on the equator, due west of Stanleyville, whose distance from Stanleyville is equal to the meridian distance of Andros island from Stanleyville.
Q is a point due east of Andros island and in latitude 38° N. Its distance from Andros island along the circle of latitude also equals the above meridian distance. Calculate the longitude of Q.

5   Find the length of the shortest great circle route from Oates Land, Antarctica (70° S, 165° W) to Amalfi (41° N 15° E).

6   Find the distance measured along the circle of latitude from Kurri Kurri in New South Wales (33° S, 151° E) to San Luis (33° S, 66° W).
Calculate the difference in the length of journey going via the South Pole, following circles of longitude instead of travelling round a parallel of latitude.

7   Sitka is in latitude 57° N and longitude 135° W, and Rizaiyeh is in latitude 38° N and longitude 45° E. Calculate
   a) the length of the great circle route from one to the other, via the North Pole
   b) the length of the route going South from Sitka along the meridian then east along the circle of latitude to Rizaiyeh.

8   X and Y are two points on the same meridian. The distance between them, measured along the meridian, is the same as the distance between two points on the equator whose longitudes are 120° W and 45° E, measured round the equator. If the position of X is 70° N, 48° W, what are the two possible positions of Y?

9   X is the point 37° N, 48° W and Y is the point 56° N, 21° W. Is it a shorter distance to travel from X to Y along the parallel 37° N and then along the meridian 21° W, or along the meridian 48° W and then along the parallel 56° N? (This question can be answered without numerical calculation.)

121

*10 Each of the vertices of a regular hexagon drawn on a plane is at an equal distance from the adjacent vertex and also from the centre of the hexagon. There are six points equally spaced round a parallel of latitude, and the distance of each from the adjacent point (measured round the parallel) is equal to its distance from the North Pole (measured along the meridian). On which parallel of latitude are the six points?

13C Areas, Volumes, Horizons

1 The area of the surface of a sphere is $4\pi R^2$ where R is the radius. The circumference of the equator is 40 000 km. What is the radius of the earth to 3 s.f.? What is the area of the surface in square kilometres? Give your answer in standard form correct to 3 s.f.

2 The total surface area of the land masses on the earth is 1.504×10^8 km². What percentage of the earth's surface is land, and what percentage water? Give your answers to 2 s.f.

3 The table gives the areas of several continents, islands and seas. Copy it into your book and add a third column giving each as a percentage of the total surface area of the earth.

	Area in km²
Australia	7.64×10^6
North America	2.43×10^7
Great Britain	2.18×10^5
Pacific Ocean	1.65×10^8
Mediterranean Sea	2.50×10^6
Cuba	1.14×10^5

4 The volume of a sphere is $\frac{4}{3}\pi R^3$, where R is the radius. Using the value of R you calculated in question 1, find the volume of the earth in km³. Write your answer in standard form correct to 3 s.f.

5 The total mass of the earth is 6.1×10^{21} tonnes. What is the average mass of 1 km³? Given that the mass of 1 km³ of water is 10^9 tonnes, how many times heavier is the earth than the same volume of water, i.e. what is the average specific gravity of the earth?

The Distance of the Horizon

*6 The diagram shows the plane of any· great circle. P is a point above the earth's surface, e.g. the top of a tower or the top of the mast of a ship. BOA is the diameter of the earth which, when extended, passes through P, and PT is a tangent through P. The distance of the horizon from P is approximately PT. From section 11B we know that $PT^2 = PA . PB$.

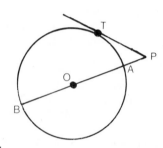

Using this formula, answer the following questions:

a) The top of the mast of a ship is 30 m above sea level. Express this in kilometres. This is PA.

b) Taking AB, the diameter of the earth, as 12 730 km, and taking PB as equal to AB, what is $PA . PB$?

c) Taking the square root of this number gives PT, the distance of the horizon in km. What is this distance?

d) The distance of the horizon should be taken as arc AT. Will there be much error in taking it as PT?

7 Using the method of question 7, find the distance of the horizon from the top of the mast of a ship   *a)* 50 m high   *b)* 60 m high.

8 Find the distance of the horizon from an aeroplane flying over the sea at the following heights above sea level:
   a) 1 km   *b)* 2 km   *c)* 5 km   *d)* 10 km

9 Make similar calculations for the distance of the horizon from the top of a tower standing in the middle of a plain and of height
   a) 40 m   *b)* 80 m   *c)* 100 m
Would you expect your answer to be accurate on land?

10 What would be the distance of the horizon from the top of Mount Everest (8848 m) at a point where land near the horizon was 1000 metres above sea level?

∗11 The formula $d = k\sqrt{h}$ gives d, the distance of the horizon in kilometres from a point h km above the sea. Find the value of k correct to 3 s.f.

14 More Algebra

14A

Factorise the following:

1. $ax^2 + bx^3 - 2cx^4$
2. $x^5 - 3x^4 + 2x^3 - x^2$
3. $x^2y^3 + 2x^3y^2 - x^4y$
4. $12a^2b^2 - 3ab^3$
5. $16a^3b^2 - 8a^2b^3 + 4a^2b^2c$
6. $x^2 - 5x$
7. $x^2 - 4$
8. $25 - 9a^2$
9. $3a^2 - 27b^2$
10. $100 - x^2$
11. $14x - 7x^2$
12. $49 - 9a^2$
13. $a^2b^2 - 1$
14. $6x^2 - 24y^2$
15. $x^2 - 3x - 28$
16. $x^2 + 8x + 15$
17. $x^2 + 12xy + 20y^2$
18. $a^2b^2 - ab - 12$
19. $a^2 + 3ab - 10b^2$
20. $6 - a - a^2$

14B

Write each of the following as a product of its factors:

1. $a^2 - 1$
2. $3a^2 - 75$
3. $64b^2 - 1$
4. $64c^2 - 16$
5. $49^2 - 21^2$
6. $9 \cdot 65^2 - 1 \cdot 35^2$
7. $119^2 - 69^2$
8. $(27 \times 119) - (7 \times 119)$
9. $(3 \cdot 6 \times 14 \cdot 5) + (3 \cdot 6 \times 5 \cdot 5)$
10. $(96 \times 13) + (48 \times 14)$

14C

Factorise:

1. $x^2 - 23x - 24$
2. $2x^2 - 16x + 24$
3. $3x^2 - 6x - 24$
4. $4x^2 - 28x + 24$
5. $2x^2 + 14x + 24$
6. $4x^2 - 20x - 24$
7. $2x^2 + 22x - 24$
8. $a^2 - 4ab - 21b^2$
9. $a^2 - 11ab + 30b^2$
10. $a^2b^2 - 9ab + 20$
11. $14 - 5x - x^2$
12. $18 + 3xy - x^2y^2$

14D

Find the factors of each of the following:

1. $6x^2 - x - 1$
2. $6x^2 + 5x - 1$
3. $6x^2 + 7x - 3$
4. $4a^2 + 5a + 1$
5. $4a^2 - a - 3$
6. $4b^2 - 4b - 15$

7 $2b^2 - b - 15$

8 $6p^2 + 5p - 6$

9 $6p^2 + 15p + 6$

10 $4q^2 + 11q + 6$

11 $3y^2 + 4y - 4$

12 $8 - 10y - 3y^2$

13 $9 - 12x - 5x^2$

14 $12 + 17x + 6x^2$

15 $2 - 7y + 3y^2$

16 $10 + 11y - 6y^2$

17 $24 + 7p - 6p^2$

18 $12p^2 - 28p + 15$

19 $12a^2 - 7a - 10$

20 $8q^2 - 17q + 9$

14E

Factorise:

1 $a(b-1) + 2(b-1)$

2 $a(x+y) - b(x+y)$

3 $x(x-2) + a(x-2)$

4 $p(p^2+1) - q(p^2+1)$

5 $p(q+p) - (q+p)$

6 $a(b-c) + c(c-b)$

7 $x(3-x) - 2(x-3)$

8 $2x^2 - 2x + y(x-1)$

9 $p^2 - pq - 3(q-p)$

10 $xy + x - x^2(1+y)$

11 $ax + ay + bx + by$

12 $x^2 - xy - 2x + 2y$

13 $a + a^2 - ab - a^2b$

14 $pq + p - q - 1$

15 $ay - xy + ax - x^2$

16 $p^2 - p + q - q^2$

17 $x^2 - x^2y + y^2 - y^3$

18 $ap - bp - a + b$

19 $a^2b - a^2 + ab - a$

20 $a^2b - b^3 - a^2 + b^2$

21 $x^2 + ax^2 - x - ax$

22 $ay^2 - by^2 - a + b$

23 $p - q - pq + q^2$

24 $x^2 - x^3 - 1 + x$

14F

1 Given that $x-1$ is a factor of $x^2 + 5x - 6$, what answer would you expect from dividing the function by $x-1$? Check this by long division.

2 Given that $x-1$ is a factor of $x^3 - 4x^2 + 2x + 1$, find the quotient when the function is divided by $x-1$.

3 Find the quotient when the first function is divided by the second in each of the following:

a) $x^2 - x - 12$, $x+3$

b) $x^2 - 6x + 8$, $x-2$

c) $x^3 - x^2 - 5x - 3$, $x+1$

d) $x^3 - 2x^2 - 9x + 4$, $x-4$

e) $x^3 - 3x^2 - 8x + 4$, $x+2$

f) $x^3 + 8x^2 + 13x - 10$, $x+5$

g) $x^3 + x^2 - 14x + 6$, $x-3$

h) $x^3 - 6x^2 + x - 6$, $x-6$

i) $x^3 + 3x - 4$, $x-1$

j) $x^3 + 2x^2 + 9$, $x+3$

4 In each part of question 3 you should have found that there was no remainder. What does this tell you about the expression you divided by? Give one value of x which makes the function $x^2 - x - 12$ in *3a* zero. (If you are not sure, think of the solution of the equation $x^2 - x - 12 = 0$.) Give a value of x which makes the function in question *3b* zero. Repeat for *3c.....j*. Check by substitution.

5 If $f: x \to x^3 + 2x^2 - 5x - 6$, find the value of $f(2)$. Hence write down a factor of $f(x)$.

6 In each of the following state whether the first function is a factor of the second or not:

a) $x + 1$, $x^2 + 3x - 4$ b) $x + 3$, $x^2 + 4x - 3$
c) $x + 4$, $3x^2 + 10x - 8$ d) $2x - 1$, $8x^2 + 2x - 3$
e) $x - 1$, $x^3 + 2x^2 + x - 2$ f) $x + 5$, $x^3 + 4x^2 - 4x + 5$
g) $x - 2$, $x^3 - 3x^2 + x - 2$ h) $x + 4$, $2x^3 + 6x^2 - 5x + 12$
i) $2x - 1$, $2x^3 + x^2 - 4x - 2$ j) $2x + 1$, $4x^3 - 6x^2 - 3$

7 State a value of x which makes $x^3 - 8 = 0$. State a factor of $x^3 - 8$. By division find the other factor.

8 Repeat question 7 for $x^3 - 1$, $x^3 + 1$, $x^3 + 8$, $x^3 - 27$, $x^3 - 1000$.

9 Using the results of question 8, find the factors of:

a) $x^3 - a^3$ b) $x^3 + y^3$ c) $x^3 - 8y^3$ d) $8a^3 - 1$ e) $a^3 - 125p^3$

10 By first finding a simple factor of $x^3 - 7x + 6$, factorise the function completely.

14G   Miscellaneous Factors

1 Factorise the following:

a) $1 - 49a^2$ b) $ax - a^2 - ab + bx$ c) $9x^2 - 3x - 2$
d) $2x^2 - 8$ e) $y^3 + 64$ f) $x^4 - 1$
g) $12 + 29x - 8x^2$ h) $12 - 15x - 18x^2$ i) $2q - pq + 2p - 4$
j) $128 - 2a^2$

2 Factorise:

a) $p^2 - q^2 + p + q$ b) $a^2 - (b + 2a)^2$ c) $(x + 2)^2 - (y - 2)^2$
d) $(3 - 2a)^2 - (1 + 3a)^2$ e) $p^2 - 4 + 3p - 6$ f) $ab - b^2 + ac - bc$
g) $9x^2 - 4(x + 1)^2$ h) $25y^2 - x^4$ i) $(x - 1)^2 + (x - 1)(x + 3)$
j) $1 - y^6$

3 Given $f: x \to x^3 - 2x^2 - x + 2$, find a value of a such that $f(a) = 0$ and hence find three factors of $f(x)$.

4 If $f(4) = 0$ where $f: x \to x^3 + ax^2 + bx - 8$, also $f(-1) = 0$, calculate the values of a and b. Find the three factors of $f(x)$.

5 If $f: x \to x^3 - x^2 + ax + b$ and $f(-3) = 0$, also $f(5) = 0$, find the values of a and b and the three factors of $f(x)$.

6 Given that $x + 2$ is a factor of $x^3 + 3x^2 + ax - 24$, find the value of a and the other two factors.

7 Divide $x^4-2x^3-16x^2+2x+15$ by $x-5$. Hence find the other three factors.

8 If $f:x\rightarrow x^4+ax^3+bx^2-x+6$ and $f(2)=0$, also $f(-3)=0$, find the values of a and b. Hence find two other values of x which make $f(x)=0$.

14H The Remainder Theorem

1 Divide x^3+2x^2-x+4 by $x+2$ and find the remainder.

2 Find the remainder when x^3-5x^2-x+2 is divided by $x-3$.

3 Find the remainder when x^4-2x^3+4x-5 is divided by $x+1$.

4 The *Remainder Theorem* states that 'when $f(x)$ is divided by $x-a$, the remainder is given by $f(a)$'. Verify this by finding the remainders in questions 1, 2 and 3 by this method.
If you want to know *why* this is so, look at question 12.

5 Use the Remainder Theorem to find the remainder when

a) x^2-3x+5 is divided by $x+1$.
b) $3x^2-4x-2$ is divided by $x-2$.
c) x^3-7x+3 is divided by $x-1$.
d) x^3+5x^2-x-3 is divided by $x+3$.
e) x^3-10x^2+5x-3 is divided by $x-4$.

6 When x^3+ax^2+bx-6 is divided by $x-2$ the remainder is 4. When it is divided by $x+3$, the remainder is 9. Calculate the values of a and b.

7 When x^4+ax^3+bx+5 is divided by $x+2$ the remainder is 45. When it is divided by $x-3$ the remainder is 20. Find the values of a and b and one factor of the function.

8 Given that $x-5$ is a factor of x^3+ax^2+3x+b and that the remainder is -8 when it is divided by $x-3$, calculate the values of a and b. Find the other two factors.

9 The Remainder Theorem is useful for finding factors of cubic (and higher) expressions. Consider x^3-2x^2-5x+6. If $x-a$ is a factor, a must be ±1, ±2, ±3 or ±6. Trying these in turn, the remainder is zero when $x=1$, so $x-1$ is a factor. What are the other factors?

10 Find the factors of:

a) $x^3+4x^2-4x-16$ b) $x^3-9x^2+26x-24$
c) $x^3-7x^2-4x+28$ d) $x^3-6x^2+5x+12$

11 In the following examples, find one factor by the Remainder Theorem and divide out to find the others (if any):

a) x^3-4x^2+5x-2 b) x^3+x-2 c) $3x^3+2x^2-6x-20$
d) $5x^3-4x^2-7x+2$

✱12 The proof of the Remainder Theorem is quite simple. $F(x)$ is any expression in x. When $F(x)$ is divided by $(x-a)$ let the quotient be $f(x)$ with a remainder R; then $F(x)=(x-a)f(x)+R$.
Put $x=a$ into this equation and so prove the theorem.

14I Fracties

1 Simplify the following:

a) $\dfrac{3x-6}{x^2-4}$ b) $\dfrac{2a+4}{2a-6}$ c) $\dfrac{3x^2-9x}{6x^2}$ d) $\dfrac{5a^2b}{5(a^2+b)}$ e) $\dfrac{x^2-x-2}{2x-4}$

f) $\dfrac{x^2-5x-6}{x^2-6x}$ g) $\dfrac{a^2-b^2}{2a+2b}$ h) $\dfrac{4-p^2}{p^2+2p-8}$ i) $\dfrac{25x^2-1}{5xy-y}$

2 Simplify:

a) $\dfrac{15xy}{x+y}\times\dfrac{x^2-y^2}{5y^2}$ b) $\dfrac{6p^2+3p}{3p-9}\times\dfrac{2pq}{4q+8pq}$

c) $\dfrac{a^2-ab}{b^2-ab}$ d) $\dfrac{a^2-b^2}{2a+b}\div\dfrac{2a-2b}{2a^2+ab}$ e) $\dfrac{x^2-5x+6}{x+2}\div\dfrac{x^2-2x-3}{x+1}$

f) $\dfrac{3y^2-12}{y^2+3y+2}\times\dfrac{y+1}{6y-24}$ g) $\dfrac{ab-ac+c^2-bc}{ab-ac-b^2+bc}$

3 Simplify:

a) $\dfrac{2}{ab}+\dfrac{3}{b}$ b) $\dfrac{4x}{y}-\dfrac{2y}{x}$ c) $\dfrac{5}{x^2}-\dfrac{2}{x^2}$ d) $\dfrac{5p}{p+q}-2$

e) $\dfrac{3+q}{q+1}+\dfrac{1+3q}{q+1}$ f) $\dfrac{3y}{x^2-y^2}-\dfrac{2}{x+y}$ g) $\dfrac{5}{2a-6b}-\dfrac{1}{9b-3a}$

h) $\dfrac{3}{x+2}+\dfrac{x-2}{x^2-3x-10}$ i) $\dfrac{3x}{2x^2-8x}-\dfrac{x+1}{x^2-5x+4}$

Write each of the following as a single fraction:

4 $\dfrac{4}{x-2}-\dfrac{7x}{x^2-4}+\dfrac{5}{x+2}$

5 $\dfrac{1}{(x-1)^2}-\dfrac{1}{2(x-1)}+\dfrac{1}{2(x+1)}$

6 $1+\dfrac{1}{(x-1)^2}+\dfrac{2}{(x-1)}$

7 $\dfrac{2}{3(x-2)}+\dfrac{1}{3(x+1)}-\dfrac{1}{x+2}$

8 $\dfrac{1}{2(x+1)^2}-\dfrac{1}{4(x+1)}+\dfrac{1}{4(x-1)}$

9 $\dfrac{3x}{x-3}-\dfrac{2x}{x-2}$

10 $\dfrac{1}{x+1}+\dfrac{1}{x-2}-\dfrac{2}{x+2}$

11 $\dfrac{1}{x-4}+\dfrac{1}{x+4}$

12 $\dfrac{9}{2(x-3)}-\dfrac{5}{2(x-1)}-\dfrac{2}{x+2}$

15 Matrices and Rotations

15A Base Vectors and Polar Co-ordinates

Base Vectors

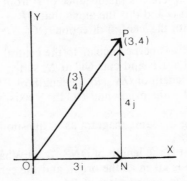

1 The vector $\begin{pmatrix} 3 \\ 4 \end{pmatrix}$ can be written in
'base vector' form as $3i + 4j$, where
i and j are the base vectors
$\begin{pmatrix} 1 \\ 0 \end{pmatrix}$ and $\begin{pmatrix} 0 \\ 1 \end{pmatrix}$ respectively.
Write the following vectors in base
vector form:

a) $\begin{pmatrix} 2 \\ 3 \end{pmatrix}$ b) $\begin{pmatrix} -1 \\ 4 \end{pmatrix}$ c) $\begin{pmatrix} 3 \\ -2 \end{pmatrix}$

d) $\begin{pmatrix} -5 \\ -1 \end{pmatrix}$ e) $\begin{pmatrix} 1 \\ -1 \end{pmatrix}$

Write the position vectors of the following points in base vector form:

f) $(4, 5)$ g) $(1, 7)$ h) $(-2, 3)$ i) $(5, -1)$ j) $(-2, -4)$

2 a) Transform the point $P(4,3)$ by the matrix $M \begin{pmatrix} 4 & 3 \\ 2 & 1 \end{pmatrix}$.

Give the co-ordinates of the image P'.

b) If i' and j' are the images of $i \begin{pmatrix} 1 \\ 0 \end{pmatrix}$ and $j \begin{pmatrix} 0 \\ 1 \end{pmatrix}$ when transformed by the

matrix M, then $i' = \begin{pmatrix} 4 \\ 2 \end{pmatrix} = 4i + 2j$. Express j' in terms of i and j in a similar way.

c) The position vector of P is $4i + 3j$. What is the value of $4i' + 3j'$ in terms of
i and j?

d) What is p' the position vector of P' in terms of i and j?

e) Comparing your answers to c and d, what do you notice?

3 Repeat question 2 using the matrices and points stated.

a) $\begin{pmatrix} 4 & 2 \\ -1 & 1 \end{pmatrix}$, $(3, 1)$. Compare p' and $3i' + j'$.

b) $\begin{pmatrix} -2 & 4 \\ 2 & 3 \end{pmatrix}$, $(1, 1)$. Compare p' and $i' + j'$.

c) $\begin{pmatrix} 1 & 1 \\ 2 & -3 \end{pmatrix}$, $(3, -2)$. Compare p' and $3i' - 2j'$.

4 If you have answered questions 2 and 3 correctly you will be able to see that if p is
the position vector $ai + bj$, and p' is the image of p under transformation by a matrix
M, then $p' = ai' + bj'$ where i' and j' are the images of i and j under the same
transformation. (The formal proof of this law is not included, but it is not difficult.)
Here are some further examples:

a) If P is the point $(2, 5)$ and P', i' and j' are the images of P, i and j

respectively under transformation by the matrix $\begin{pmatrix} 1 & -1 \\ 0 & 2 \end{pmatrix}$, write down

i) P' ii) i' iii) j' iv) $2i' + 5j'$.

b) Comparing a)i) and a)iv) is it true to say that $p' = 2i' + 5j'$?

5 Repeat question 4 with the matrix $\begin{pmatrix} 3 & 1 \\ 4 & -2 \end{pmatrix}$ and the point $(-2, 1)$.

6 Repeat question 4 with the matrix $\begin{pmatrix} -2 & 1 \\ 3 & 2 \end{pmatrix}$ and the point $(1, -2)$.

Polar Co-ordinates

7 P is the point whose polar co-ordinates are (r, θ) where r is the distance of P from the origin O and θ is the angle that OP makes with the 'central direction' OC.

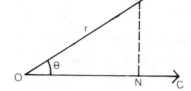

a) If PN is perpendicular to the central direction OC and cuts OC at N, state
i) the length of ON ii) the length of PN.
b) If O is the origin and OC the x axis, what are the Cartesian co-ordinates of P?

8 Using the same diagram as in question 7, if P is the point whose polar co-ordinates are $(4, 60°)$,
a) what is the length of ON?   b) What is the length of PN (to 3 s.f.)?
c) If O is taken as the origin and the central direction as the x axis, what are the Cartesian co-ordinates of P?

Note   In questions 9 to 13, O is the origin for both polar co-ordinates and Cartesian co-ordinates, and the central direction for polar co-ordinates is also the x axis for Cartesian co-ordinates.

9 What are the Cartesian co-ordinates of the following points whose polar co-ordinates are given?

a) $(3, 45°)$ b) $(7, 30°)$ c) $(4, 150°)$ d) $(2, 230°)$ e) $(3, 330°)$

10 If P is the point whose Cartesian co-ordinates are $(5, 4)$ and PN is perpendicular to OX,

a) what is the length of OP?
b) what is the value of $\tan \theta$?
c) what is θ (to the nearest degree)?
d) what are the polar co-ordinates of P?

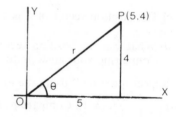

11 Repeat question 10 for the point $(4, 8)$.

12 Find the polar co-ordinates of the following points whose Cartesian co-ordinates are given: a) $(2, 3)$ b) $(1, 4)$ c) $(4, 1)$ d) $(3, -2)$ e) $(-3, 2)$.

13 If the polar co-ordinates of P are (r, θ) then the position vector OP can easily be written down in base vector form. This $(4, 60°)$ is $(4\cos 60°)i + (4\sin 60°)j$, or $2i + 3\cdot46j$. Write in base vector form the position vectors of the following points whose polar co-ordinates are given: a) $(1, 40°)$ b) $(2, 80°)$ c) $(1, 10°)$ d) $(3, 130°)$ e) $(2, 260°)$

15B   Rotations

Rotations using Polar Co-ordinates

1 P is the point $(4, 30°)$. P is rotated $40°$ about the origin to P'.

a) What is the length of i) OP ii) OP'?
b) What is i) $\angle POC$ ii) $\angle P'OC$?
c) What are the polar co-ordinates of P'?

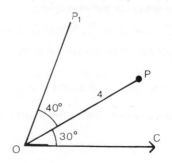

2 You are given a point P and a rotation R about the origin. In each case state the image of P under the rotation R.

a) $(3, 20°), 30°$ b) $(5, 70°), 120°$ c) $(6, 80°), -40°$ d) $(2, 120°), 130°$ e) $(r, a), b$

Rotations using Matrices

3 a) OI is the base vector i, $\begin{pmatrix} 1 \\ 0 \end{pmatrix}$.

It is rotated through $30°$ about the origin and its new position is OI'. $I'N$ is drawn perpendicular to OX. State the lengths of
i) OI ii) OI' iii) ON iv) $I'N$.
  b) What are the co-ordinates of I'?
  c) OI', or i', is the transform of i. Express i' in terms of i and j.
  d) By carrying out a similar calculation, express j' in terms of i and j.
  e) Using a similar method again, work out the values of i' and j' for a rotation about the origin through angle a.

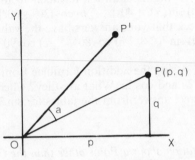

4 Now consider the point P, which is (p, q). $\overline{OP}$ is $pi+qj$. If P is transformed to P' by a rotation about the origin through angle a, then $\overline{OP'}$ will be $pi'+qj'$ (from 15A, question 4).

  a) Using the values of i' and j' you found in question 3e, express the vector OP' in terms of i and j.
  b) What are the co-ordinates of P'?

5 If you answered questions 3 and 4 correctly, you will now know that when the point (p, q) is rotated about the origin through angle a, its image is $(p\cos a - q\sin a, p\sin a + q\cos a)$.

We now have to find the matrix M such that $M\begin{pmatrix} p \\ q \end{pmatrix} = \begin{pmatrix} p\cos a - q\sin a \\ p\sin a + q\cos a \end{pmatrix}$.

Clearly $M = \begin{pmatrix} \cos a & -\sin a \\ \sin a & \cos a \end{pmatrix}$.

M is the matrix which gives a rotation about the origin through angle a.

  a) Putting a equal to i) $90°$ ii) $180°$ iii) $270°$, find the matrices which represent rotations about O through these angles.
  b) Do the matrices you found agree with the ones you already know for a quarter turn, half turn and three-quarter turn about the origin?

6 Using the matrix M from question 5, find the images of the following points under the rotations stated, giving no more than 3 s.f. in your answers:

a) $(3, 2), 30°$ b) $(4, 1), 40°$ c) $(-2, 3), 70°$ d) $(-1, -3), 50°$
e) $(0, 4), 120°$ f) $(5, 0), 140°$ g) $(3, 3), 45°$ h) $(-2, 5), 220°$
i) $(2, -3), 60°$ j) $(4, -4), -45°$

7 I is the image of the point P when it is rotated about the origin through an angle of $40°$. I_1 is the image of I when it is rotated about the origin through an angle of $50°$. I_2 is the image of P when it is rotated about the origin through an angle of $90°$. If P is the point $(1, 2)$, work out the values of I, I_1 and I_2. What do you notice about I_1 and I_2? Why is this?

131

8 Write down the matrix M_1 which represents a rotation of 32° about the origin. Write down M_2 and M_3 which represent similar rotations of 43° and 75° respectively. Work out the products M_1M_2 and M_2M_1 and compare your answers a) with each other and b) with M_3. What do you notice? What can you deduce?

The Addition Formulae for Sine and Cosine

∗9 a) Write down the matrices representing rotations about the origin of i) angle a ii) angle b iii) angle $(a+b)$. Call the matrices M_1, M_2 and M_3 respectively.
b) Work out carefully the product M_1M_2. This represents a rotation of $(a+b)$ about the origin, and should therefore be congruent to M_3.
Equate corresponding terms and you should get as your results:

$$\cos(a+b) = \cos a \cos b - \sin a \sin b$$
$$\sin(a+b) = \sin a \cos b + \cos a \sin b$$

These two formulae are known as the addition formulae.

∗10 Using the addition formulae from question 9, find to 3 s.f. the value of
a) $\sin(30° + 40°)$ b) $\cos(20° + 60°)$ c) $\cos(10° + 80°)$
Check that your answers have the values given in the tables for
a) $\sin 70°$ b) $\cos 80°$ c) $\cos 90°$.

∗11 Using the addition formulae from question 9 and putting $b=a$, gives formulae for $\sin 2a$ and $\cos 2a$. What are they? Check your formulae for $a=15°$. See if they give the correct answer from the tables for $\sin 30°$ and $\cos 30°$.

Rotation about a Point other than the Origin

∗∗12 Find the image I of the point Q (4, 3) when it is rotated through 40° about A (2, 2).

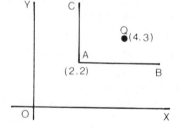

a) With A as origin, draw new axes AB and AC parallel to OX and OY. What are the co-ordinates of the object point Q with respect to these new axes?
b) Using the correct matrix, rotate Q about the new origin through 40°. Find the new co-ordinates of the image I correct to 3 s.f.
c) Mark I on your diagram and calculate the co-ordinates of I referred to the original axes OX and OY. This is the final answer.

∗13 Using the method of question 12, find the image of:
a) (5, 3) when rotated through 50° about (2, 1).
b) (2, 4) when rotated through 60° about (1, 3).
Give your answers to 3 s.f.

∗14 Repeat question 13 for the following:
a) (5, 2) rotated through 17° about (3, 4).
b) (2, 1) rotated through 71° about (4, 2).
c) (4, 3) rotated through 121° about (2, 2).

6 Probability

16A Revision and the Use of Set Notation

1 Two dice are thrown together. The diagram shows all possible scores. What is the chance of getting

a) at least one four, *b)* one and only one four, *c)* a total of four, *d)* no fours?
e) What are the odds against getting a four?

	1	2	3	4	5	6
1	*	*	*	*	*	*
2	*	*	*	*	*	*
3	*	*	*	*	*	*
4	*	*	*	*	*	*
5	*	*	*	*	*	*
6	*	*	*	*	*	*

2 A small deck of cards is made up of the king, queen, jack, ace, two, three and four of diamonds. A similar deck is made up of hearts. One card is drawn at random from each pack, the picture cards scoring 0 and the other cards scoring their face value.

a) Draw a diagram similar to the one in question 1 but with seven rows and seven columns, headed 0, 0, 0, 1, 2, 3, 4.
b) What is the chance of getting *i)* a score of 0, *ii)* a score of 3, *iii)* a score of 5, *iv)* the maximum possible score? What is this score?

3 Repeat question 2 but using all thirteen cards in each suit. How many extra rows and columns do you need to add to the diagram to get the answers?

4 Using the same decks of cards as in question 2 but counting each diamond as worth twice is face value and each heart three times its face value, what is the probability of a score of

a) 0, *b)* 7, *c)* 11, *d)* 15, *e)* the maximum possible score? What is this score?

(*Hint* Label the rows of your diagram 0, 0, 0, 2, 4, 6, 8 and the columns 0, 0, 0, 3, 6, 9, 12.)

5 Answer question 4 if you subtract the smaller face value from the larger.

6 The set of possible values in each of questions 1 to 5 is known as the 'sample space' or 'possibility space' and is denoted by $\mathscr{E}$. In question 1, $n(\mathscr{E})$ is 36. What is $n(\mathscr{E})$ in question 2 and question 3?

7 In question 1, a total score of 5 could have been obtained as (4, 1), (3, 2), (2, 3) or (1, 4). If we call a score of 5 'event *a*' and denote by *A* the set of all possible ways of getting *a*, then $A = \{(4, 1), (3, 2), (2, 3), (1, 4)\}$ and $n(A) = 4$. $p(a)$, the probability of *a*, is then $\dfrac{n(A)}{n(\mathscr{E})}$.

Write down *X* in full and also $n(X)$, $n(\mathscr{E})$ and $p(x)$ in each of the following cases:

a) *x* is a score of 4 in question 1.   *b)* *x* is a score of 4 in question 2.
c) *x* is a score of 6 in question 4.

8 In triangle *PQR*, $\angle P$ is an integral multiple of $9°$, $\angle Q$ of $10°$ and $\angle R$ of $12°$.

a) Draw up a table giving all possible values of *P*, *Q* and *R*. It could start as shown.

P	18°	18°
Q	150°	90°
R	12°	72°

b) These sets of possible values constitute $\mathscr{E}$, the possibility space. What is $n(\mathscr{E})$?
c) Write down $n(Y)$ and $p(y)$ where y is the event stated in each of the following cases: *i*) the triangle has an angle of 18°, *ii*) the triangle has an angle of 36°, *iii*) the triangle has an angle of 72°, *iv*) the triangle is isosceles, *v*) the triangle is right-angled.
d) If z is the event 'the triangle has an angle of $x°$' and $p(z)$ is $\frac{4}{11}$, what is the value of x?

9 A bird-watcher notices that a large bird table in his garden is regularly visited by four magpies, two male (M_1 and M_2) and two female (F_1 and F_2). Sometimes there is one magpie on the table, sometimes two, but the table is not big enough for more than two at a time.

a) List the members of $\mathscr{E}$, the sample space, i.e. all possible combinations of magpies on the table at one given time. Your list could contain M_1, M_2, M_1F_1, etc.
b) Considering only the times when there are one or more magpies on the table, if all the events in $\mathscr{E}$ are equally likely, and if the following numbers of birds on the table are denoted by the letters stated: one bird, q; two birds, r; two male birds, s; two female birds, t; one male and one female bird, u; three birds, v; state *i*) $n(\mathscr{E})$, *ii*) $n(Q)$, $p(q)$, *iii*) $n(R)$, $p(r)$, *iv*) $n(S)$, $p(s)$, *v*) $n(T)$, $p(t)$, *vi*) $n(U)$, $p(u)$, *vii*) $n(V)$, $p(v)$.

10 *a*) Denoting heads by H and tails by T, write down the sample space $\mathscr{E}$ when four tenpence pieces are tossed together. If could start $\{HHHH, HHHT, \dots\}$
b) What is $n(\mathscr{E})$?
c) If the following events are represented by the letters stated: three heads, m; one head, n; equal numbers of heads and tails, q; state *i*) $p(m)$, *ii*) $p(n)$, *iii*) $p(q)$.

11 A party of eight cyclists on a hilly moorland road are riding in pairs, but the pairs are constantly changing as the cyclists tackle the hills at their own speeds. Three bicycles are red (R_1, R_2 and R_3), two green (G_1 and G_2) and three blue (B_1, B_2 and B_3).
a) Write down the sample space $\mathscr{E}$ showing all possible combinations of bicycles in pairs.
b) What is $n(\mathscr{E})$?
c) If all members of $\mathscr{E}$ are equally likely, using set notation or otherwise, state the probabilities of a pair consisting of *i*) two red bicycles, *ii*) one red and one green, *iii*) two bicycles of the same colour, *iv*) two bicycles of different colours.

12 Two dice are thrown together. What is the probability that
a) the number on one die is a factor of the number on the other die?
b) the numbers on the two dice have a common factor?
Take 1 as a factor in both parts of the question.

13 *a*) Write down the sample space $\mathscr{E}$ when 3 coins are tossed together.
b) What is $n(\mathscr{E})$?
c) What is the probability of *i*) no heads, *ii*) 1 head, *iii*) 2 heads, *iv*) 3 heads? (Give your answers with $n(\mathscr{E})$ in the denominator. Do not cancel down.)

14 Repeat question 13 for 4 coins, arranging the sample space in columns and not using set brackets. Part *c* must include '*v*) 4 heads'.

15 Repeat question 14 for 5 coins. Part *c* must include '*vi*) 5 heads'.

16 In questions 13, 14, 15, do you recognise the numbers in *a*) the numerators of the fractions, *b*) the denominators of the fractions?

17 Using the results of question 16, write down the probabilities of:
  a) 4 heads when 7 coins are tossed together.
  b) 3 heads when 8 coins are tossed together.
  c) 7 heads when 9 coins are tossed together.

16B Combining Probabilities

1 In a group of ten boys, six have bicycles and eight have sisters. If you were asked to find the probability of a boy having both a bicycle and a sister, one way of finding an answer would be to consider the 'possibility space' shown in the diagram. A nought represents a boy with a sister, a cross a boy with a bicycle, a cross over a nought a boy with both, and a dot a boy with neither.

BICYCLES

SISTERS

a) There are 100 possibilities altogether. In how many of these does a boy have both a bicycle and a sister?

b) 'A boy has a bicycle and a sister' is a combined event. 'A boy has a bicycle' and 'a boy has a sister' are two independent, separate events. What is the probability of *i*) the first of these separate events, *ii*) the second of these separate events?

c) Is the probability of the combined event, calculated in *a*, the same as the product of the probabilities of the two separate events, calculated in *b*?

2 If you answered question 1 correctly, you will now know that the probability of a combined event, when the two separate events are independent, is the product of the probabilities of the two separate events. This is much quicker than drawing a diagram of the possibility space. Now answer the following question:
Isobel goes to the library with three tickets. The chances of finding a book by one of her three favourite authors are $\frac{1}{2}$, $\frac{3}{4}$ and $\frac{1}{3}$ respectively. What is the chance that she will find three books, one by each author?

3 In the month of June the chance of a very hot day (over 30°C) is $\frac{1}{15}$, and the chance of a chilly day (under 15°C) is $\frac{1}{6}$. A supermarket has one or more dairy products on special offer four days out of six. On a certain weekday in June, what will be the probability of dairy products being on offer *and a*) the day being very hot, *b*) the day being chilly?

4 Peter cycles to school every day along a rough road. The chance of heavy rain on the way to or from school is 1 in 40 and the chance of a puncture is 1 in 60. What is the chance of a puncture in heavy rain?

5 On average Paul arrives late for school once in six weeks (30 school days). He forgets something such as dinner money, homework or P.E. kit on average one day in ten. On a certain day, what is the probability that he will both have arrived late and also have forgotten something? (Assume the two events are independent. In actual practice they may not be.)

6 Judith and Deborah have four cats, Bimbam, Twitten, Blackie and Gobolina. They also have three dogs, Josie, Scampy and Shaggy. When the animals 'want in' the cats scratch at one door, and the dogs at another.

 a) If there is one cat scratching, what is the probability that it is Twitten?
 b) If there is one dog scratching, what is the probability that it is Scampy?
 c) If one cat and one dog are scratching at the same time, what is the probability that it is i) Twitten and Scampy, ii) Gobolina and Josie?

7 Sometimes two events are mutually exclusive. For instance, if I throw two dice and they show a total of 10, then they cannot at the same time show a total of 5. The diagram shows these two mutually exclusive events.

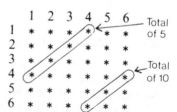

 a) What is the probability of the first (i.e. a total of 10)?
 b) What is the probability of the second (i.e. a total of 5)?
 c) What is the probability of either the first or the second?
 d) Is the probability in c the sum of the probabilities in a and b?

8 In a week of seven days, Mrs Philpot cooks meat for her family on three days and fish on two days. On a given Wednesday, what is the chance that she will be cooking either meat or fish?

9 If you pick a card at random from a full deck of playing cards, what is the chance of it being a) a picture card   b) an ace   c) either a picture card or an ace?

10 A soldier is on sentry duty one day in five and has chicken for dinner four times in 30 days. What is the probability that on a given day he will either be on sentry duty or have chicken for dinner?

11 Stephanie goes roller-skating on one weekday every week, and she takes the dog for a walk on four of the *other* weekdays. What is the chance on a given weekday that she will either go roller-skating or take the dog for a walk?

12 30% of the staff in a city office walk to work and 50% come by car. If an individual member of staff is selected at random, what is the chance that they either walk to work or come by car?

13 Events are not always independent or mutually exclusive.

 a) When two dice are thrown together, what is the probability of a double?
 b) What is the probability of a score of eight?
 c) What is the probability of either a score of eight or a double or both?
 d) Is the probability in c the sum of the probabilities in a and b?

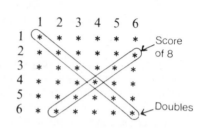

In your answers to each part of questions 14–19, insert one of the letters I, M or N to show that the separate events are independent, mutually exclusive or neither.

14 In a Rugger fifteen, ten of the players have long hair and nine are tall. All the players are equally liable to get a sprained ankle. If one player sprains an ankle, what is the chance that this player is short and has short hair?

15 In autumn, I go into the fields looking for blackberries and mushrooms. If the chance of finding blackberries is $\frac{3}{4}$ and the chance of finding mushrooms is $\frac{2}{7}$,

   a) what is the chance that I find *i)* both blackberries and mushrooms
   ii) neither   *iii)* either blackberries or mushrooms but not both?
   b) What is the sum of your three answers? Why?
   c) Draw a diagram illustrating the 'possibility space', and shade each of the areas corresponding to *a i), ii)* and *iii)* with distinctive shading. (One area will be divided into two parts.)

16 I throw three coins together. What is the chance of *a)* the first showing a head *b)* the second showing a head   *c)* the third showing a head *d)* three heads *e)* three tails *f)* neither three heads nor three tails?

17 In the month of June, the chance of a very hot day (i.e. over 30°C) is $\frac{1}{15}$, and the chance of a chilly day (i.e. under 15°C) is $\frac{1}{6}$. What is the chance of a certain day being *a)* either very hot or very chilly *b)* both very hot and very chilly *c)* neither?

18 In a certain restaurant customers can dine 'à la carte', i.e. they can order any items that are on the menu. If one out of four order soup, four out of five order fish, and one out of two order ice cream, what is the chance of a customer ordering *a)* all three *b)* both the last two *c)* only the last two (consider the second, third and 'not first') *d)* either the second only or the first only?

19 *a)* When one die is thrown, what is the chance of *i)* a four   *ii)* not a four?
   b) When three dice are thrown together, what is the chance of *i)* three fours
   ii) no fours (i.e. three 'non fours')   *iii)* a four on the first die only (i.e. non fours on the other two)   *iv)* a four on any one die but not on the other two
   v) a four on each of the first and second dice but not on the third   *vi)* a four on any two dice, but not on the other one?
   c) What is the sum of the probabilities in *i), ii), iv)* and *vi)* of *b*? Why?

16C Miscellaneous Probability

Note In two of questions 1–13 the information given is insufficient to calculate an answer.

1 The probability tree shown opposite can be used to calculate the probabilities of getting 2's with three throws of a die. For each throw the probability of 2 is $\frac{1}{6}$ and of 'not 2' (represented by N) $\frac{5}{6}$. Multiply probabilities along the branches and add them across the branches.

   a) What is the probability of three 2's? (Look at the branch marked A.)
   b) What is the probability of two 2's? (Look at the branches marked B.)
   c) What is the probability of one 2? (Look at the branches marked C.)
   d) What is the probability of no 2's? (Look at the branch marked D.)

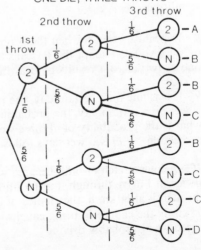

ONE DIE, THREE THROWS

e) What is the sum of the probabilities in *a*, *b*, *c* and *d*? Why?

f) What is the probability of 2 on the second throw only?

g) What is the probability of 'not 2' on both the first and second throws?

2   A coin is weighted so that the probability of a head is $\frac{2}{3}$ and of a tail $\frac{1}{3}$.

a) Draw a tree diagram to illustrate four throws of the coin. (You will find it easiest to start with a column of *H*, *T*, *H*, *T*,... spaced evenly down the right-hand side of the paper, and to build up the tree from right to left.)

b) From your diagram find the probabilities of *i*) four heads   *ii*) three heads
iii) two heads   *iv*) one head   *v*) no heads.

c) What is the sum of the probabilities in *b*? Why?

3   Four coins are thrown together once only. One is a fair coin, one is weighted as in question 2, and the other two are weighted so that the chance of a head is $\frac{3}{5}$. Draw a tree diagram and answer the same questions as in 2*b* and *c*. The diagram will be similar to that in question 1, but the headings will be *Four coins, one throw* and *1st coin, 2nd coin*, etc.

4   A school dinner consists of soup, fish or lamb, and ice cream or jam pudding. The school supervisor estimates that the probability of a pupil having soup is $\frac{1}{2}$. If they choose soup, the probability of them choosing fish is $\frac{9}{10}$ and if they do not choose soup the probability of them choosing lamb is $\frac{2}{3}$. If they choose fish, the probability of them choosing ice cream is $\frac{7}{9}$, and if they choose lamb the probability of them choosing jam pudding is also $\frac{7}{9}$. Draw a tree diagram and find *a*) the probability of ice cream being chosen   *b*) the probability of jam pudding being chosen. Assume that no pupil refuses either of the last two courses.

5   When the striker of a football team makes a shot, the probability of scoring a goal is $\frac{1}{2}$. If he makes four shots in the first ten minutes of a game, what is the probability of him scoring 4, 3, 2, 1 or 0 goals respectively? Work the answer *a*) by using Pascal's triangle   *b*) by drawing a tree diagram.

6   Repeat question 5 for six shots, working the answer for 6, 5, 4, 3,...0 goals, but do not attempt the tree diagram unless you have a large sheet of paper.

*7   Repeat question 5 when the probability of scoring is $\frac{2}{3}$. Draw the tree diagram first and with its help see if you can adapt the Pascal's triangle solution to fit the new conditions.

**8   Repeat question 6 when the probability is $\frac{2}{3}$. Use your experience from question 7 to see if you can adapt the Pascal's triangle solution to fit the new conditions. Check one or two of your results by drawing the relevant branches of the tree diagram.

9   If it is fine on a certain day, the probability of the next day being wet is $\frac{1}{5}$, but if it is wet on a certain day, the probability of the next day being wet is $\frac{3}{5}$. After a fine day, what is the probability of *a*) three wet days in succession   *b*) four wet days in succession   *c*) two wet days followed by a fine day?

10   A long stretch of built-up area with a 50 km/h speed limit is regularly patrolled by police. If I drive through it at 65 km/h, the chance of being caught is $\frac{1}{60}$. If I drive through it at 80 km/h, the chance of being caught is $\frac{1}{10}$. If I drive through it at 95 km/h, the chance of being caught is $\frac{1}{3}$. If I am caught twice in one week, at what speed was I probably driving?

11 The Pennine Walk is one of the longest footpaths in Great Britain, extending from Edale in Derbyshire to Kirk Yetholm in Midlothian. If a party completes the whole walk of 435 kilometres, the chance of them doing so without a minor accident is $\frac{49}{50}$. The chance of them doing so without at least one thorough soaking is $\frac{1}{50}$. The chance of losing the way at least once is $\frac{7}{10}$. What is the chance of a party completing the walk without either a minor accident, a thorough soaking or losing the way at least once? Give your answer as a decimal correct to 3 s.f.

12 The probability of a boy of sixteen being over 1·8 metres tall is $\frac{1}{100}$, the chance of him weighing more than 90 kg is $\frac{1}{200}$, and the chance of him wearing size 11 shoes is $\frac{1}{250}$. If a boy of sixteen is selected at random, what is the chance of him being over 1·8 metres tall, weighing over 90 kilograms and taking size 11 in shoes?

13 The chance of a boy of sixteen playing cricket for his school is $\frac{1}{10}$, the chance of him taking private lessons in music is $\frac{1}{100}$, and the chance of him speaking both French and German is $\frac{1}{500}$. If Tom is sixteen, what is the probability of him doing
a) all three   *b)* none of the three? Give your answer to *a* in standard form correct to 2 s.f. and your answer to *b* as a decimal correct to 3 d.p.

Miscellaneous Examples B

B1

1 a) Find the factors of each of the following: i) $4x^2y+6xy^2$ ii) $4a^2-b^2$
iii) $1-8p^3$ iv) $2x^2-7x+5$ v) $n+nm+m+m^2$
  b) Solve the equation $\dfrac{2x}{3}-\dfrac{3(x-2)}{5}=1$.

2 a) Change the binary number 101 110 into base 4 and into base 8.
  b) Change the base 3 number 11 212 into base 9.
  c) Change the number 305 which is in base n, where n is greater than 5, into base n^2.

3 In the diagram BR and PC are parallel,
and BC and PQ are parallel.
  a) If $AR=2$ cm, $RC=3$ cm and $AB=4$ cm,
calculate the lengths AP and AQ.
  b) If the area of triangle ABR is x sq. units,
what is i) the area of triangle CBR
ii) the area of triangle APC iii) the
area of quadrilateral $BRCP$?

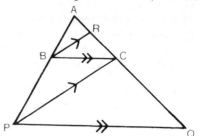

4 Nine cards, the 2, 3, 4, 5, 6, 7, 8, 9 and 10 of clubs, are lying face down on the table.

  a) If I pick one up what is the probability that it is the four?
  b) If the first card is the four and I pick up a second card without replacing the first,
what is the probability that it is the five?
  c) When all the cards are on the table, I pick up a pair. What is the probability that
they are consecutive numbers?
  d) With all the cards on the table, I pick up three. What is the probability of these
three being consecutive numbers?

5 a) If $f:x\to4x-3$ and $g:x\to2x^2$, find: i) $f(-1)$ ii) $g(-3)$ iii) $f^{-1}(x)$
iv) $gf(x)$ v) values of y for which $gf(y)=50$.
  ✱b) If $F:n\to a$ where a and n are both positive integers and a is the highest prime
factor of n, find the values of: i) $F(12)$ ii) $F(21)$ iii) $F(25)$
iv) What do you know about n if $F(n)=n$?
v) Is the statement $FF(n)=a$ true for all values of n?

✱*6* a) Calculate, in nautical miles, the great circle distance between Acapulco ($17°$ N,
$100°$ W) and Virdu ($50°$ N, $100°$ W).
  b) Calculate in nautical miles the length of the arc of the circle of latitude $30°$ N,
which subtends an angle of one minute at the centre of this circle. Hence find the
distance between Agadir ($30°$ N, $10°$ W) and New Orleans ($30°$ N, $90°$ W).

B2

✱*1* The table shows three sets of values of x and y
which are two related variables.

x	0	1	2
y	0	2	n

  a) If y is a linear function of x, find n.
  b) If $n=8$, write down the equation which relates x and y.
  c) If the first pair of values are excluded and y is inversely proportional to x, find n.
  d) If the first pair of values are excluded, $n=\sqrt{2}$ and y is inversely proportional to
x^p, what is the value of p?

2 Two teams each with four members took part in an inter-school quiz. Their individual scores for the first three rounds are shown in the table.

Team members	A	B	C	D	P	Q	R	S
Scores Round 1	7	5	8	3	4	6	3	7
Round 2	5	3	7	4	5	8	3	5
Round 3	5	6	9	6	5	8	4	8

a) Calculate the mean score and the median score for each round. Which round did the teams find the simplest?
b) Write down the average score per round of the member *i)* with the highest individual score   *ii)* with the lowest individual score.
c) After the fourth round, *B* had an average of 5. What was *B*'s score in round 4?
d) After round 4, *Q*'s average score had dropped by $\frac{1}{3}$ of a point. What did *Q* score in that round?
e) After the fourth round, *P* had the same average score as *R* and *S* would have had if their scores had been put together at the end of round 3. What did *P* score in round 4?

3 *ABCDEFGH* is an octagon formed in such a way that the four squares shown in the diagram are each of side 2 cm. Describe the following transformations fully:

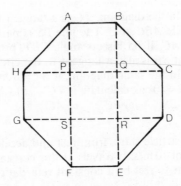

a) an enlargement that maps triangle *APH* on to triangle *FPC*
b) a translation followed by a rotation that maps triangle *APH* on to triangle *RQC*
c) a reflection that maps triangle *FPC* on to triangle *GRB*
d) a transformation that maps triangle *ADE* on to triangle *EHA*.

4 Two similar-shaped containers of varying circular cross-sections hold 125 cm³ and 1 litre of liquid respectively. If the smaller one has a maximum radius of 4 cm, what is the maximum radius of the larger one? Write down the ratio of the areas of their bases.

5 Values of *x* and *y* are governed by the following inequalities:
$x+y<5$, $y<3x$ and $2y>x-2$. State whether the following statements must be true or may be true: *a)* $y<4$ *b)* $x<4$ *c)* $y<x$ *d)* $x>0$ *e)* $y>-2$.

6 A rectangular piece of card, whose length is twice its width, has a square of side 2 cm cut from each corner. The sides are then folded up to form an open rectangular box of depth 2 cm. Show that the volume of this box is $4(x-4)(x-2)$ where *x* is the width of the original card in cm. If the volume is 320 cm, find the length and width of the original card.

B3

1 *a)* Write down the perimeter of the given shape.
b) Write down an expression for the area of the shape.
c) If the area is 78 cm² and $y=3$ cm, find the value of *x*.

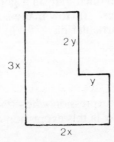

2 $x = \dfrac{k\sqrt{y}}{z^2}$ where k is constant.

a) If $x = 2$ when $y = 36$ and $z = 3$, find k.
b) If x is kept constant, state how y varies with z.

3 P and Q are two places on the circle of
latitude $30°$ N, at opposite ends of a
diameter of this circle.

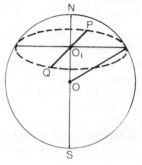

a) Find the distance from P to Q via the
North Pole. Take the radius of the earth
as 6370 km.
b) Express the distance over the pole as a
fraction of the distance from P to Q round
the circle of latitude.

4 a) In the diagram, TC is a tangent to
circle ABC at C. The line BS is parallel
to AC. If BS bisects angle CBT, prove
that SB is also a tangent to the circle.
b) In the same diagram, if $BT = 8$ cm,
$AB = 10$ cm, calculate i) TC ii) CS
iii) BS iv) BC v) AC.

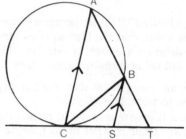

5 A particle starts from rest and accelerates uniformly for 4 seconds to reach a
speed of 10 m/s. It travels at this constant speed for the next $2\frac{1}{2}$ seconds, and is then
brought to rest by a constant retardation of 2 m/s^2.

a) Draw a velocity/time graph for this particle.
b) Calculate its acceleration during the first 4 seconds.
c) Calculate the distance it travels during the whole period of motion.
d) Find its maximum speed in km/h.

6 The table shows the number of eggs of each grade collected by a poultry farmer in
one day.

Grade	7	6	5	4	3	2	1
Weight in grams	up to 45	45–50	50–55	55–60	60–65	65–70	70+
Frequency	13	24	42	48	39	23	11

Draw a cumulative frequency graph to show this information.
a) From your graph find   i) the median weight (in what grade does the median
lie?)   ii) the semi-interquartile range   iii) the percentage of the day's eggs which
weighed 53.5 g or less.
b) The farmer reckoned that on a good day, 20% of the eggs collected would be in
the top two grades.   i) How many short of this figure were the eggs in these two
grades on this particular day?   ii) What percentage of the total did this represent?

B4

1 Write down the expression whose factors are x, $x - 5$ and $x + 2$.
For what values of x is this expression equal to 6?

2 The points $A(1, 0)$, $B(2, 0)$ and $C(1, 2)$ form a triangle T_1.
$A'(-3, 1)$, $B'(-1, 1)$ and $C'(-3, 5)$ form triangle T_2.
a) Describe the transformation which maps T_1 on to T_2.
b) Triangle T_1 is then enlarged by a scale factor -2 with centre $(0, 1)$ on to $A''B''C''$
(triangle T_3). Find the co-ordinates of A'', B'' and C''. Describe the transformation
which maps T_3 on to T_2.

3 John, Neil, David, Lynne, Sarah and John's sister Kate form a tennis team.

a) When they play mixed doubles, what is the probability that John plays with his
sister?
b) When they play men's doubles, John's brother Simon joins them. What is the
probability that John will not be paired with his brother?
c) From the seven players two pairs are picked to represent the club in a mixed
doubles match. What is the probability of Kate being chosen? What is the
probability of Kate being chosen and her being partnered by one of her brothers?

4 A model of a factory is made on the scale of 1 to 500.

a) The length of the model is 1·2 m. Calculate the corresponding length of the factory
in km.
b) Part of the factory stands on an area of 75 hectares. What is the corresponding
area on the model?

5 In the diagram TD is the tangent to
circle $ABCD$ at D. AB is parallel to DT.
DC meets AB produced at E.

a) Prove that triangles AED and CAD are
similar and that $AD^2 = CD \cdot DE$.
b) Prove that triangles CAD and CEB are
similar.
c) If $CE = 4$ cm, $DC = 5$ cm and
$BC = 2·5$ cm, find the length AC.

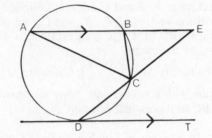

6 On a globe the distance between two points A and B on the equator is 37·7 cm. A
is $38°$ E and B is $110°$ E.

a) Find the radius of the globe.
b) If C is due north of B, a distance of 23·6 cm measured on the surface of the globe,
what is the position of C?
c) D is on the same circle of latitude as C at $160°$ W. Find the shorter distance
between C and D measured along the parallel of latitude.

B5

1 Simplify a) $\dfrac{3}{4} + \dfrac{4}{3}$ b) $\dfrac{x}{y} + \dfrac{y}{x}$ c) $\dfrac{x-2}{x+2} + \dfrac{x+2}{x-2}$

2 If A and B are any two subsets of a given universal set $\mathscr{E}$, $A°B$ is defined to mean
the set $A \cap B'$.

a) Draw a Venn diagram to illustrate the set $A°B$.
b) If $\mathscr{E} = \{$Positive whole numbers from 1 to 20 inclusive$\}$, $A = \{$Even numbers$\}$,
$B = \{$Multiples of 3$\}$ and $C = \{1, 2, 3, 4, 5, 6, 7, 8, 9, 10\}$, list the sets $(A°B)°C$ and
$A°(B°C)$.
(OXFORD)

3 If it is fine I go to school on my bicycle. If it is wet I go on the bus. The probability of it being wet is $\frac{1}{5}$. If I go to school on my bicycle the probability of being late is $\frac{1}{4}$, but if I go on the bus the probability of being late is $\frac{2}{3}$. Find the probability of

a) going to school on my bicycle
b) going on my bicycle and being late
c) arriving late for school.

If I go to school on the bus, I come home on the bus. The probability of my missing the bus and being late home is $\frac{2}{5}$. The probability of being late home if I am on my bicycle is $\frac{1}{6}$. Find the probability of

d) coming home on the bus
e) arriving home late on the bus
f) arriving home late on my bicycle
g) arriving home late on a fine day when I also arrived at school late.

4 A, B and C are the three points whose position vectors are $\begin{pmatrix} 1 \\ 0 \end{pmatrix}$, $\begin{pmatrix} 1 \\ 3 \end{pmatrix}$ and $\begin{pmatrix} 2 \\ 0 \end{pmatrix}$ respectively. Two matrices P and Q are given by $P = \begin{pmatrix} 0 & -1 \\ 1 & 0 \end{pmatrix}$ and $Q = \begin{pmatrix} 1 & 0 \\ 0 & -1 \end{pmatrix}$.

Show on graph paper the triangle ABC and its image $A_1B_1C_1$ when the position vectors of A, B and C are premultiplied by P. Show also the image $A_2B_2C_2$ when the position vectors of A_1, B_1 and C_1 are premultiplied by Q. State the single matrix which maps ABC to $A_2B_2C_2$, and state also what single geometrical transformation this matrix represents.

The matrix $M = \begin{pmatrix} 0 & 1 \\ -1 & 1 \end{pmatrix}$. Calculate M^2 and show that $M^3 = P^2$.

State with a reason how many successive multiplications by the matrix M will restore ABC to its original position. (OXFORD)

5 In a survey of London traffic, 500 drivers were asked to state the length of their journey. Their replies are recorded in the following table:

Length of journey (km)	0–4	4–8	8–12	12–16	16–20	20–24	24–28	28–32
Frequency	35	50	104	108	90	62	41	10

a) Calculate the mean journey length from these data.
b) Construct a cumulative frequency table, and draw the cumulative frequency curve.
c) From your curve estimate the interquartile range. (MEI)

6 A lampshade is part of a sphere of radius 12 cm. Using a similar reference system to that used for the earth's surface, the rim of the bottom is the circle of latitude 55° S and the top is the circle of latitude 80° N.
On the top is a cylindrical collar of height 2 cm. A spider crawls from the bottom rim to the top and then starts to make a web inside by leaving a vertical thread hanging. Calculate:

a) the distance the spider has crawled if it came the shortest way
b) the length of thread hanging inside the shade if it just reaches from the top to the bottom rim
c) the distance of the hanging thread from the nearest point on the bottom rim.

17 Surds

17A

Square roots can often be simplified, e.g. $\sqrt{18}=\sqrt{9\times 2}=\sqrt{9}\times\sqrt{2}=3\sqrt{2}$.

Simplify the following, making the number under the square root sign as small as possible:

$1 \quad \sqrt{8}$	$2 \quad \sqrt{18}$	$3 \quad \sqrt{50}$	$4 \quad \sqrt{75}$	$5 \quad \sqrt{48}$
$6 \quad \sqrt{98}$	$7 \quad \sqrt{45}$	$8 \quad \sqrt{54}$	$9 \quad \sqrt{242}$	$10 \quad \sqrt{700}$
$11 \quad \sqrt{28}$	$12 \quad \sqrt{52}$	$13 \quad \sqrt{175}$	$14 \quad \sqrt{72}$	$15 \quad \sqrt{128}$

17B

Given that $\sqrt{2}=1\cdot41$, $\sqrt{3}=1\cdot73$ and $\sqrt{5}=2\cdot24$, find the values of each of the following:

$1 \quad \sqrt{12}$	$2 \quad \sqrt{27}$	$3 \quad \sqrt{8}$	$4 \quad \sqrt{20}$	$5 \quad \sqrt{18}$
$6 \quad \sqrt{32}$	$7 \quad \sqrt{80}$	$8 \quad \sqrt{162}$	$9 \quad \sqrt{180}$	$10 \quad \sqrt{108}$

17C

Remembering that $\sqrt{2}\times\sqrt{2}=2$, $\sqrt{3}\times\sqrt{3}=3$, etc., simplify the following, leaving your answer in surd or integral form:

$1 \quad \sqrt{12}+\sqrt{27}$	$2 \quad \sqrt{2}+\sqrt{8}$	$3 \quad \sqrt{75}-\sqrt{50}$	$4 \quad \sqrt{80}-\sqrt{20}$
$5 \quad 3\sqrt{20}+2\sqrt{45}$	$6 \quad \sqrt{175}-5\sqrt{7}$	$7 \quad \sqrt{147}-\sqrt{108}$	$8 \quad 3\sqrt{32}-2\sqrt{18}$
$9 \quad \sqrt{5}\times\sqrt{20}$	$10 \quad \sqrt{2}\times\sqrt{18}$	$11 \quad 5\sqrt{2}\times3\sqrt{2}$	$12 \quad \sqrt{6}\times3\sqrt{2}$
$13 \quad \sqrt{8}\times\sqrt{18}$	$14 \quad 3\sqrt{5}\times\sqrt{10}$	$15 \quad \sqrt{75}\times\sqrt{12}$	$16 \quad \sqrt{288}\times\sqrt{162}$
$17 \quad \sqrt{125}\times\sqrt{500}$	$18 \quad 3\sqrt{8}\times2\sqrt{27}$	$19 \quad \dfrac{\sqrt{6}}{\sqrt{3}}$	$20 \quad \dfrac{\sqrt{21}}{\sqrt{7}}$
$21 \quad \dfrac{\sqrt{96}}{2\sqrt{6}}$	$22 \quad \dfrac{\sqrt{32}}{\sqrt{8}}$	$23 \quad \dfrac{4\sqrt{5}}{\sqrt{45}}$	$24 \quad \dfrac{\sqrt{128}}{\sqrt{8}}$
$25 \quad \dfrac{\sqrt{128}}{\sqrt{32}}$	$26 \quad (\sqrt{2})^3$	$27 \quad (2\sqrt{2})^2$	$28 \quad (\sqrt{3})^3$
$29 \quad (\sqrt{3})^4$	$30 \quad (5\sqrt{3})^2$		

17D

Remembering that $\dfrac{3}{\sqrt{3}}=\sqrt{3}$, $\dfrac{2}{\sqrt{2}}=\sqrt{2}$, etc., rationalise the following, i.e. make all the denominators rational numbers.

Examples: $\dfrac{6}{\sqrt{3}}=\dfrac{2\times 3}{\sqrt{3}}=2\times\sqrt{3}$; $\dfrac{1}{\sqrt{5}}=\dfrac{\sqrt{5}}{\sqrt{5}\times\sqrt{5}}=\dfrac{\sqrt{5}}{5}$;

$$\dfrac{1}{\sqrt{3}+1}=\dfrac{\sqrt{3}-1}{(\sqrt{3}+1)(\sqrt{3}-1)}=\dfrac{\sqrt{3}-1}{3-1}=\dfrac{\sqrt{3}-1}{2}$$

1 $\dfrac{4}{\sqrt{2}}$ *2* $\dfrac{1}{\sqrt{2}}$ *3* $\dfrac{\sqrt{12}}{\sqrt{3}}$ *4* $\dfrac{2}{\sqrt{3}}$ *5* $\dfrac{2}{\sqrt{5}}$

6 $\dfrac{3}{\sqrt{6}}$ *7* $\dfrac{9}{\sqrt{15}}$ *8* $\dfrac{3\sqrt{7}}{\sqrt{21}}$ *9* $\dfrac{4\sqrt{5}}{\sqrt{20}}$ *10* $\dfrac{\sqrt{8}+\sqrt{2}}{\sqrt{12}}$

11 $\dfrac{1}{\sqrt{2}-1}$ *12* $\dfrac{1}{\sqrt{3}+\sqrt{2}}$ *13* $\dfrac{1}{\sqrt{3}-\sqrt{2}}$ *14* $\dfrac{1}{\sqrt{5}-\sqrt{3}}$ *15* $\dfrac{2}{\sqrt{5}+2}$

17E

Given that $\sqrt{2}=1\cdot41$, $\sqrt{3}=1\cdot73$ and $\sqrt{5}=2\cdot24$, calculate the values of each of the following, correct to 2 decimal places. To simplify your calculations, first make all the denominators rational.

1 $\dfrac{1}{\sqrt{2}}$ *2* $\dfrac{1}{\sqrt{3}}$ *3* $\dfrac{1}{\sqrt{5}}$ *4* $\dfrac{1}{\sqrt{12}}$ *5* $\dfrac{3}{\sqrt{18}}$

6 $\dfrac{15}{\sqrt{200}}$ *7* $\dfrac{12}{\sqrt{80}}$ *8* $\dfrac{3}{\sqrt{3}-1}$ *9* $\dfrac{\sqrt{5}}{\sqrt{5}-2}$ *10* $\dfrac{\sqrt{2}}{2+\sqrt{2}}$

17F   Surds in Geometry and Trigonometry

Where appropriate, leave your answers to these questions in surds.

1   Calculate the length of the diagonal of a square of side 4 cm.

2   Repeat question 1 for squares with sides of 5 cm, 7 cm, 10 cm, 60 cm and p cm.

3   For each of the squares in question 2, if O is the intersection of the diagonals and A is any vertex of the square, find the length of OA.

4   A right pyramid has a square base of side 4 cm and a perpendicular height of 4 cm. Find the length of the slant edges.

5   The perpendicular height of a right pyramid, on a square base of side 8 cm, is 4 cm. Calculate the length of the slant edges.

6   Find the length of the diagonal of a cuboid 8 cm by 6 cm by 5 cm.

7   Find the length of the diagonal of a cuboid with dimensions 6 cm, 3 cm and 2 cm.

8 Four squares each with sides of 2 cm are joined as shown in the diagram. An octagon *ABCDEFGH* can be drawn by joining the vertices. Calculate

    a) the perimeter of the octagon
    b) the distance between *AB* and *EF*
    c) the distance between *BC* and *FG*.
    d) Are all the angles of the octagon equal?
    Is it a regular octagon?

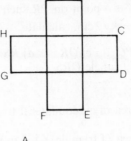

9 The diagram shows triangle *ABC* in which angle *B*=90°, angle *C*=45° and *BC*=1 unit. Write down the lengths of *AB* and *AC*.
Hence write down the sine, cosine and tangent of 45°, leaving your answer in fractions.

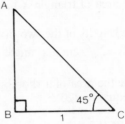

10 *PQR* is an equilateral triangle with sides of 2 units. *M* is the mid-point of *QR*. Find the lengths of *QM* and *PM*, and using triangle *PQM* write down the sine, cosine and tangent of 30° and of 60°. Leave your answers in fractions.

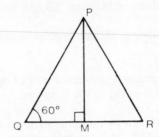

For questions 11 to 22 use the ratios for 30°, 45° and 60° which you have just found.

11 Calculate the lengths *a*, *b*, ... *h* in the following diagrams.

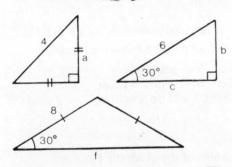

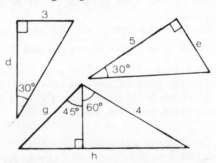

12 The sides of a rectangle are $\sqrt{12}$ cm and 2 cm. Find the angle between the diagonals.

13 In the diagram, triangle *ABC* has angle *B*=90°, angle *A*=30° and *AC*=5 cm.
Calculate the length of *BC* and *BD*, and hence the area of the triangle.
There are other methods of calculating the area of this triangle. Find them and check that your answer is the same in each case.

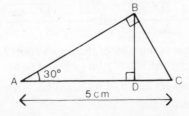

14 In triangle PQR, angle $P=90°$ and angle $Q=60°$. S is a point on PR such that $PS=PQ$. If $QS=3\sqrt{2}$ cm, calculate

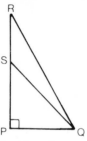

a) PS b) PQ c) QR d) RS
e) the area of triangle QRS.

15 Find the area of an equilateral triangle with sides of 6 cm.

16 Find the area of triangle XYZ in which $XY=XZ=8$ cm and angle $X=135°$.

17 Find the lengths of the two diagonals of the parallelogram $PQRS$ in which $PQ=4$ cm, $PS=2\sqrt{3}$ cm and angle $P=30°$.

18 Calculate the area of a rhombus which has sides of 10 cm and one pair of opposite angles each equal to 120°.

∗19 $ABCDEFGH$ is a regular octagon with sides of 6 cm. Calculate
  a) the distance between AB and EF b) BG c) AC d) AE.

20 The length AD in a regular hexagon $ABCDEF$ is 18 cm. Calculate the area of the hexagon.

21 The distance between a pair of opposite sides in a regular hexagon is $4\sqrt{3}$ cm. Calculate the area of the hexagon.

22 $VABC$ is a pyramid on a triangular base ABC in which $AB=AC=12$ cm and angle $A=120°$. M is the mid-point of BC and O is the point in AM such that $2MO=OA$. V is vertically above O at a height of 4 cm. Calculate

a) the slant edge VA b) the slant edge VB c) the area of the base
d) the volume of the pyramid   e) the area of VBC.

23 The sides of a triangle are a, b and c and s is the semi-perimeter, $\frac{1}{2}(a+b+c)$. Use the formula $\sqrt{s(s-a)(s-b)(s-c)}$ to find the area of each of the following triangles:

a) $a=4$ cm, $b=7$ cm, $c=9$ cm   b) $a=5$ cm, $b=7$ cm, $c=6$ cm
c) $a=7$ cm, $b=9$ cm, $c=8$ cm   d) $a=7$ cm, $b=11$ cm, $c=12$ cm
e) $a=11$ cm, $b=15$ cm, $c=16$ cm   f) $a=19$ cm, $b=11$ cm, $c=10$ cm
g) $a=13$ cm, $b=15$ cm, $c=22$ cm.

24 *The Roots of a Quadratic Equation*
In Chapter 7 we gave two useful checks for the roots of the quadratic equation $ax^2+bx+c=0$. The sum of the roots should be $-b/a$ and the product of the roots should be c/a.
You are now in a position to verify these checks. As you already know the roots are $\dfrac{-b+\sqrt{b^2-4ac}}{2a}$ and $\dfrac{-b-\sqrt{b^2-4ac}}{2a}$.

a) What is the sum of these two roots?
b) Using the identity $(m+n)(m-n)=m^2-n^2$, what is the product of these two roots?

Your answers to a and b should be $-b/a$ and c/a respectively and this verifies the check.

8 Vectors

18A Revision

1 If **a** is a vector and *n* is a scalar, describe *n**a***.

2 Is vector addition commutative, i.e. is it true to say that **a**+**b**=**b**+**a**? Give diagrams to illustrate your answer.

3 Is vector addition associative, i.e. is it true to say that
(**a**+**b**)+**c**=**a**+(**b**+**c**)=**a**+**b**+**c**?

4 Is it true to say that multiplication by a scalar is distributive over vector addition, i.e. that *n*(**a**+**b**+**c**)=*n**a***+*n**b***+*n**c***?

5 What is the length of the column vector $\begin{pmatrix} x \\ y \end{pmatrix}$? What is the tangent of the angle the vector makes with the *x* axis?

6 If **a**=*x**i***+*y**j*** where **i** and **j** are base vectors, give an expression for |**a**|, the length of **a**, in terms of *x* and *y*.

7 Each of the following sets of diagrams illustrates one of the 'laws' in questions 1–6. State which law.

a)

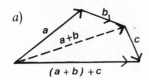 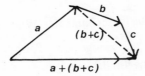 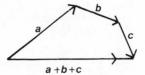

b)

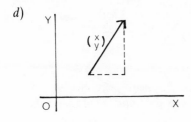 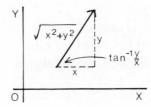

c)

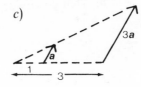

d)

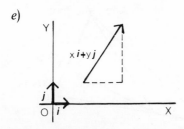 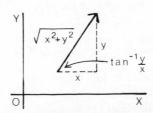

e)

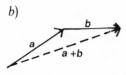 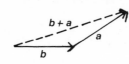

149

f)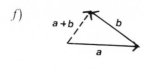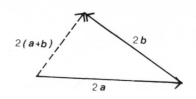

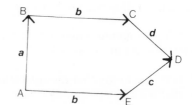

8 *a)* Using the diagram opposite, give vector expressions for

i) $\overline{AD}$ ii) $\overline{AC}$ iii) $\overline{BE}$ iv) $\overline{CE}$
v) $\overline{BD}$ vi) $\overline{CA}$

b) There are two possible answers to each of the above questions. Selecting any pair, equate them and see what you find.

9 Using the diagram from question 8, simplify: *a)* $b+c-d$ *b)* $\overline{AE}+\overline{BC}$
c) $a-c+d$ *d)* $2c-d-a$ *e)* $a+b-c+d$

10 *a)* From triangle *BCD* find an expression for b in terms of a and c.
b) From triangle *BEA* find an expression for $\overline{EA}$ in terms of a and c.
c) What do you deduce about *EA* and *CD*?
d) Find vector expressions for $\overline{EC}$ and $\overline{AD}$. What do you deduce about *EC* and *AD*?

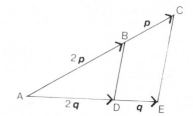

11 *a)* Find vector expressions for $\overline{BD}$ and $\overline{CE}$.
b) What do you deduce about *BD* and *CE*?

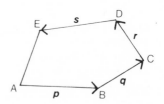

12 *a)* Find vector expressions for i) $\overline{BD}$, ii) $\overline{BE}$, iii) $\overline{CE}$, iv) $\overline{CA}$, v) $\overline{DA}$, vi) $\overline{EA}$.
b) Are there any parallel lines in the figure?

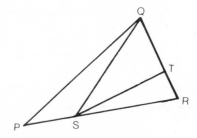

13 In the diagram, $PS=\frac{1}{2}SR$, $RT=\frac{1}{2}TQ$, $\overline{PQ}=a$, $\overline{PR}=b$.

a) Find i) $\overline{PS}$, ii) $\overline{QS}$, iii) $\overline{SR}$, iv) $\overline{QR}$, v) $\overline{RT}$, vi) $\overline{ST}$, vii) $\overline{TQ}$.
b) Using your answers to ii), vi) and vii), does $\overline{ST}$ equal $\overline{SQ}+\overline{QT}$?
c) Name any parallel lines in the figure.

14 a) Find i) $\overline{PR}$, ii) $\overline{PS}$, iii) $\overline{RS}$.
   b) If O is the mid-point of TQ, find
   i) $\overline{OR}$, ii) $\overline{OS}$.

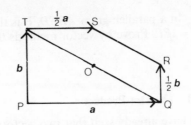

15 In the diagram opposite, give vector
expressions for:
   a) $\overline{AC}$ b) $\overline{AD}$ c) $\overline{EB}$ d) $\overline{EC}$ e) $\overline{DB}$

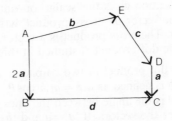

✱16 Square $OLMN$ has $\overline{OL}=a$, $\overline{ON}=b$,
$OP=3PN$ and $LQ=3QM$.
   a) If PS and TQ are parallel, $OT=\frac{1}{4}OL$
   and $NS=x(NM)$, find x.
   b) If $NS=y(NM)$ and PS and OQ are
   parallel, find y.

✱17 a) Using the diagram in question 15,
   name two pairs of parallel lines in the
   figure if $d=c-2a$.
   b) If $b+a=c$, name two pairs of parallel
   lines in the figure.
   c) If $c=a$, what can you say about b and d? Why?
   d) If $d=2(c-a)$, name two pairs of parallel lines in the figure.

✱18 In the diagram, $\overline{AC}=b$ and $\overline{AB}=c$.
   a) Give vector expressions for
   i) $\overline{AE}$, ii) $\overline{AD}$, iii) $\overline{ED}$.
   b) Express $\overline{CB}$ in terms of b and c.
   c) Comparing your answers to a and b,
   what can you deduce about
   i) the lines ED and CB
   ii) the vectors $\overline{ED}$ and $\overline{CB}$?

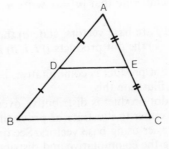

✱19 $ABCDEF$ is a hexagon.
   a) If $|a|=|b|=|c|$, is $ABCDEF$ a regular
   hexagon? Draw a figure to illustrate your
   answer.
   b) What *additional* condition do you need
   to make $ABCDEF$ regular?

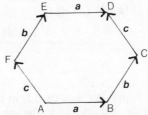

20 In a parallelogram *ABCD*, *E* is the mid-point of *CD*. *BE* is produced to *F* so that *BF*=2*BE*. Prove by vector methods that *F* lies on *AD* produced and *DF*=*AD*.

18B The Dot Product

We have already seen that two vectors *a* and *b* in the same plane can be 'added' and 'subtracted'. They can also be 'multiplied' and this can be done in two ways. Scalar multiplication gives the 'scalar' or 'dot' product, written *a*.*b*. Vector multiplication gives the vector or 'cross' product, written *a*×*b*. The dot product is a scalar, i.e. a number. The cross product is a vector, in a plane perpendicular to the original plane. Only the dot product is studied in this book.

1 The dot product of two co-planar vectors *a* and *b* is defined as $a.b=|a||b|\cos\theta$, where θ is the angle between the lines of action or the vectors. If $|a|=a$ and $|b|=b$ then the dot product is $ab\cos\theta$.

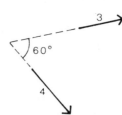

Thus if *a* is a vector of length 3 and *b* is a vector of length 4, and the angle between their lines of action is 60°, as shown in the diagram, then *a*.*b* is $12\cos 60°$, i.e. 6.

In the following examples you are given the length of each of a pair of vectors and the angle each makes with the *x* axis. Find the dot product of each pair.

  a) 3, 20°; 4, 34°   *b)* 5, 10°; 2, 40°   *c)* 6, 57°; 3, 28°
  d) 2, 86°; 1, 4°   *e)* 8, 17°; 2, 73°

✷ 2 Vector *a* joins the points (2, 4) and (3, 6). Vector *b* joins the points (5, 1) and (7, 3).

  a) Find the length of each vector (to 3 s.f.) and the angle it makes with the *x* axis (correct to the nearest tenth of a degree).
  b) Find *a*.*b* correct to 3 s.f.

3 *a)* If *a* and *b* are perpendicular vectors, what is *a*.*b*?
  b) If *a* and *b* are parallel vectors, what is *a*.*b*?
  c) Remembering that $|a|=a$, write *a*.*a* in its simplest form.

4 If *i* and *j* are base vectors, state *a)* the length of each *b)* the angle each makes with the *x* axis *c)* the dot products *i)* *i*.*i*, *ii)* *i*.*j*, *iii)* *j*.*i*, *iv)* *j*.*j*.

5 *a)* The dot product is commutative, i.e. *a*.*b*=*b*.*a*. Make up an example of your own to illustrate this.
  b) The dot product is distributive over addition, i.e. *a*.(*b*+*c*)=*a*.*b*+*a*.*c*. It is tedious either to illustrate or prove this by the methods we have used so far, but much easier using base vectors. See question 11.
  c) Using the commutative and distributive properties we can simplify expressions such as (*a*+*b*).(*c*+*d*).

$$(a+b).(c+d)=(a+b).c+(a+b).d$$
$$=c.(a+b)+d.(a+b)$$
$$=c.a+c.b+d.a+d.b$$
$$=a.c+b.c+a.d+b.d$$

State which law was used *i)* in the 1st line of the above proof, *ii)* in the 2nd line, *iii)* in the 3rd line, *iv)* in the 4th line.

6 Work out the product $(p+q).(r-s)$ giving every step in full and stating which law was used for each step (as in question 5c).

7 Give the final answer only for the following dot products, leaving out the intermediate steps.

a) $(2c+d).(e+f)$ b) $(9g-h).(2m-n)$ c) $(a+b).(a-b)$.
d) $(2a+b).(c-2d)$ given that a and c are perpendicular.
e) $(a+b).(a-c)$ given that b and c are parallel.
f) $(i+2j).(2i+j)$ where i and j are base vectors.
g) $(2i-j).(3i+j)$.

8 The dot product of column vectors can be obtained by first writing them in base vector form.

a) If $a=\begin{pmatrix}3\\4\end{pmatrix}$, write a in base vector form.

b) If $b=\begin{pmatrix}2\\5\end{pmatrix}$, write b in base vector form.

c) Write $a.b$ in base vector notation and evaluate the product.

9 Find the dot product of the following pairs of vectors:

a) $\begin{pmatrix}4\\3\end{pmatrix}$ and $\begin{pmatrix}5\\2\end{pmatrix}$ b) $\begin{pmatrix}3\\4\end{pmatrix}$ and $\begin{pmatrix}5\\2\end{pmatrix}$ c) $\begin{pmatrix}2\\7\end{pmatrix}$ and $\begin{pmatrix}-2\\4\end{pmatrix}$

d) $\begin{pmatrix}1\\-8\end{pmatrix}$ and $\begin{pmatrix}2\\2\end{pmatrix}$ e) $\begin{pmatrix}5\\6\end{pmatrix}$ and $\begin{pmatrix}4\\-5\end{pmatrix}$ f) $\begin{pmatrix}x_1\\y_1\end{pmatrix}$ and $\begin{pmatrix}x_2\\y_2\end{pmatrix}$

10 a) If you answered question 2, write the vectors in that question in column form and find their product using the method of questions 8 and 9.
b) Does your answer agree with the answer in question 2?
c) Which is the more likely to be wrong?

11 We are now in a position to illustrate quite simply that the dot product is distributive over addition.
Let $a=\begin{pmatrix}2\\3\end{pmatrix}$, $b=\begin{pmatrix}3\\1\end{pmatrix}$, $c=\begin{pmatrix}4\\2\end{pmatrix}$ and $d=b+c$.

a) Write d as a column vector.   b) Evaluate $a.d$.   c) Evaluate $a.b$ and $a.c$.
d) Does $a.(b+c)=a.b+a.c$?

✳12 Give a formal proof that the dot product is distributive over addition by using the method of question 11 but starting with $a=\begin{pmatrix}x_1\\y_1\end{pmatrix}$, $b=\begin{pmatrix}x_2\\y_2\end{pmatrix}$, $c=\begin{pmatrix}x_3\\y_3\end{pmatrix}$.

✳13 'I buy m pencils at a cost of p pence each and n pens at a cost of q pence each. Express the total cost as the product of a row vector and a column vector and find this product.'
There were a great many questions of this type in an earlier book.

The answer is of course $(m\ \ n)\begin{pmatrix}p\\q\end{pmatrix}=mp+nq$.

a) We have already mentioned two kinds of vector multiplication in this chapter. Is this a third kind?
b) Explain why the first vector was written as a row and the second as a column.

18C   Vectors and Conventional Geometry

Many theorems and riders from conventional geometry can be proved very easily using vector methods. (In some cases, however, the vector proof is not easier than the conventional proof.)

1   The line joining the mid-points of the sides of a triangle is parallel to the base and equal in length to half the base.

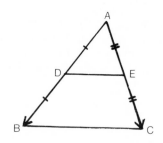

 a) In triangle ABC, $\overline{AB}$ is c and $\overline{AC}$ is b. Give a vector expression for $\overline{BC}$.

 b) D and E are the mid-points of AB and AC respectively. Give vector expressions for i) AD, ii) AE, iii) DE.

 c) Comparing a and b iii) proves the theorem.

2   If F is the mid-point of BC in question 1, and F is joined to D and E, prove that the four small triangles in the figure are all congruent.

3   Prove by vector methods that if two opposite sides of a quadrilateral are equal and parallel, then the quadrilateral is a parallelogram.

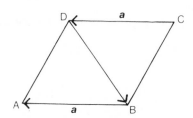

4   Prove by vector methods that if the diagonals of a quadrilateral bisect each other, the quadrilateral is a parallelogram.

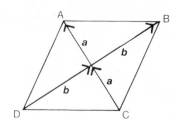

5   If two points divide the sides of a triangle in the same ratio, use vectors to prove that the line joining these points is parallel to the base. Let D and E divide AB and AC respectively in the ratio $n:1$, so that $AD = n.AB$ and $AE = n.AC$. Let $\overline{AB} = c$ and $\overline{AC} = b$. Find expressions for DE and BC from triangles ADE and ABC. Compare these expressions.

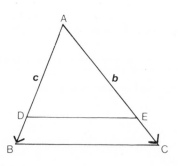

6   *Pythagoras Theorem*

ABC is a triangle right-angled at C. $\overline{AB} = c$, $\overline{BC} = a$, and $\overline{AC} = b$.
Expressing c in terms of a and b gives $c = a + b$.

 a) Using the dot product, $c.c = (a+b).(a+b)$, simplify both sides as much as possible.

 b) Does this prove the theorem?

7 Prove that the line joining the vertex of an isosceles triangle to the mid-point of the base is perpendicular to the base.

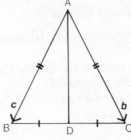

a) Let $\overline{AB}=c$, $\overline{AC}=b$. What can you say about b and c?

b) Give a vector expression for
i) $\overline{BC}$, ii) $\overline{BD}$, iii) $\overline{AD}$

c) What is the value of the dot product $\overline{BD}.\overline{AD}$? Can you prove it is zero?

d) Does this prove the theorem?

8   *The Cosine Rule*
From the figure, $c=a-b$.
Hence $c.c=(a-b).(a-b)$.
Simplify both sides as much as possible, and get the cosine rule.

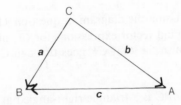

9   *The Theorem of Apollonius*
Prove that if D is the mid-point of the base of any triangle ABC, then
$AB^2+AC^2=2AD^2+2BD^2$.

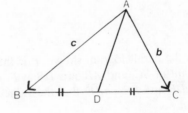

a) If $\overline{AB}=c$ and $\overline{AC}=b$, express i) $\overline{BC}$, ii) $\overline{BD}$, iii) $\overline{AD}$ in terms of b and c.

b) Write down $\overline{BD}.\overline{BD}+\overline{AD}.\overline{AD}$ in terms of b and c.

c) Simplify the expression in b. Does this prove the theorem?

10   Vector solutions to problems in geometry can often be obtained quite simply by using the following theorem.
If $ma=nb$, where m and n are scalars, then *either* a and b are parallel or collinear *or*, if a and b are not parallel or collinear, $m=n=0$. Thus to prove the diagonals of a parallelogram bisect each other,
let $\overline{DA}=a$ and $\overline{DC}=b$,
let $\overline{DP}=m\overline{DB}=m(a+b)$,
let $\overline{AP}=n\overline{AC}=n(b-a)$,
then since $\overline{DP}+\overline{PA}=\overline{DA}$,
$m(a+b)-n(b-a)=a$.
Rearranging this, $a(m+n-1)=b(n-m)$.
But a and b are not parallel or collinear,
so $m+n-1=0$ and $n-m=0$.

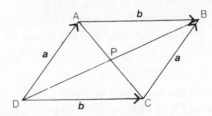

a) Solve these equations and find the values of m and n.

b) Does this prove that the diagonals of a parallelogram bisect each other?

∗11 The medians AD and BE of a triangle intersect at G.
Let $\overline{GD}=m\overline{AD}$ and $\overline{GE}=n\overline{BE}$.

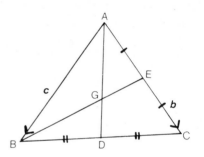

    a) Find expressions for i) $\overline{AD}$, ii) $\overline{GD}$, iii) $\overline{BE}$, iv) $\overline{GE}$.

    b) Find the value of $\overline{GC}$ i) from triangle GDC, ii) from triangle GEC.

    c) Equate your two values of $\overline{GC}$, and find the values of m and n.

∗12 Using the diagram in question 11, if CG produced cuts AB at F, prove that F is the mid-point of AB, i.e. that the three medians meet at G.

∗13 Using the diagram in question 11 again, if F is the mid-point of AB join GC and CF. Find vector expressions for $\overline{GC}$ and $\overline{GF}$ and hence show that GC and GF are collinear, i.e. that CF goes through G and the three medians of the triangle all meet at G.

14 ABC is a triangle right-angled at A. If D is the mid-point of BC, give a vector proof that $AD=\frac{1}{2}BC$.

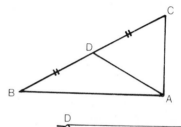

15 In the parallelogram shown, E is the mid-point of BC, and AE cuts DB at F. If $\overline{FB}=m\overline{DB}$ and $\overline{FE}=n\overline{AE}$, find the values of m and n.

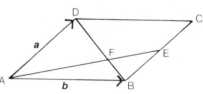

16 If in the diagram in question 15, $\overline{BE}=m\overline{BC}$, $\overline{DF}=\frac{3}{4}\overline{DB}$ and $\overline{AF}=n\overline{AE}$, find the values of m and n.

∗17 In triangle ABC, D divides BC in the ratio $BD:DC=3:2$ and E divides CA in the ratio $CE:EA=2:1$. AD and BE cut at G. Let $\overline{GD}=m\overline{AD}$ and $\overline{GE}=n\overline{BE}$. Using the method of question 11, find two expressions for $\overline{GC}$. Equate them and find the values of m and n.

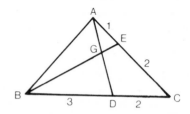

∗18 Using the figure and the results of question 17, if F divides AB in the ratio $AF:FB=1:3$, *a)* what is $\overline{AG}$? *b)* from triangle AFG, what is $\overline{FG}$? *c)* what is $\overline{GC}$? *d)* comparing $\overline{GF}$ and $\overline{CG}$, is CGF a straight line?

19 Repeat questions 17 and 18 with $BD:DC=1:2$, $CE:EA=1:2$, $AF:FB=4:1$.

156

20 The transversal *PQRST* cuts the sides of a triangle *ABC* (produced if necessary) at *Q*, *R* and *S* as shown in the figure. If $AQ:QB=3:2$ and $BR:RC=1:2$, prove that $CS:SA=4:3$.

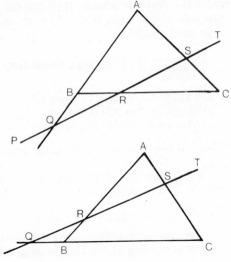

21 Repeat question 20 for the figure shown, given that $AR:RB=3:1$ and $BQ:QC=1:4$. Find $CS:SA$.

18D Vectors and Physical Quantities

Velocities

1 I walk along the deck of a small ship at 2 m/s and the ship is moving in the same direction at 3 m/s. If I had stayed still on the deck, in one second I would have moved from A_1 to A_2. But I moved an extra two metres forward in that second, so my true position is A_3. What is my true velocity?

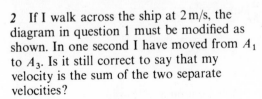

2 If I walk across the ship at 2 m/s, the diagram in question 1 must be modified as shown. In one second I have moved from A_1 to A_3. Is it still correct to say that my velocity is the sum of the two separate velocities?

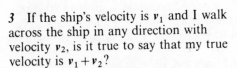

3 If the ship's velocity is v_1 and I walk across the ship in any direction with velocity v_2, is it true to say that my true velocity is v_1+v_2?

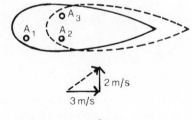

4 *The Triangle of Velocities*
A ship sails due north at 10 km/h. It is carried off course by a current running northeast at 4 km/h. These velocities are represented by the vectors $\overline{AB}$ and $\overline{BC}$ in the diagram (not drawn to scale). The 'resultant' velocity is $\overline{AC}$. Draw the diagram to scale and measure the magnitude and direction of the resultant velocity.
A diagram of this type is called a 'triangle of velocities'.

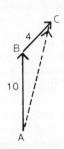

157

5 In the previous triangle BAC use the cosine rule to find the length of AC. Use the sine rule to find angle BAC. Do your measured results agree reasonably well with the calculated results?

6 Use either the cosine rule or accurate scale drawing to combine the following pairs of velocities:

    a) 6 km/h due W and 11 km/h NW   *b)* 90 km/h on bearing 200°, 70 km/h on 160°
    c) 25 m/s on 050°, 34 m/s on 086°   *d)* 35 km/h on 063°, 12 km/h on 153°
    e) 14 m/s on 053°, 17 m/s on 356°.

Check at least one answer by using both methods.

7 A ship steaming at 27 km/h on a bearing of 280° is carried off course by a current running at 4 km/h on a bearing 045°. What is the true speed of the ship and on what bearing?

8 The captain of a ship sailing at 20 km/h wishes to sail due N, but a current from the SW is running at 3 km/h. On what course should he steer to sail due N?

9 An aeroplane flying due E at an air speed of 500 km/h to an aerodrome at which it wishes to land is blown off course by a steady wind of 60 km/h from the NE. If the destination of the aircraft is now on a bearing 280°, what course should the captain set so as to arrive at his destination without further change of course?

10 An object which is slightly heavier than water falls off a ship in a deep lagoon and sinks slowly at 1 m/s. If at that point there is a current running at 0·5 m/s, at what angle to the vertical does the object sink, and at what speed does it move?

11 Pairs of velocities are represented by the vectors stated. Find their resultants.

    a) $\binom{3}{2}$ and $\binom{4}{1}$ *b)* $2i+3j, 4i-j$ *c)* $\binom{-2}{5}$ and $\binom{1}{-1}$
    d) $-2i-3j, 3i+j$ *e)* $\binom{2}{-4}$ and $\binom{-4}{1}$

Relative Velocity

12 Peter is in a suburban train travelling out of London at 80 km/h.
Paul is in a main line train on a parallel track, leaving London at a speed of 90 km/h. Peter spots Paul and waves to him. Through the carriage window Paul appears to be moving slowly forwards. Paul spots Peter and waves to him. Peter appears through the carriage window to be moving slowly backwards.
The velocity of Paul relative to Peter is 10 km/h forward, or $+10$ km/h.
What is the velocity of Peter relative to Paul?

13 State first the velocity of A relative to B, then the velocity of B relative to A in each of the following cases:

    a) A moving forward at 30 km/h, B moving forward at 60 km/h.
    b) A moving forward at 20 m/s, B moving backward at 20 m/s.
    c) A at rest, B moving backward at 16 km/h.
    d) A moving back at 40 m/s, B moving forward at 10 m/s.

14 *a)* Look at your eight answers to question 13. Is it right to say that to get the velocity of A relative to B you add the velocity of A to the reversed velocity of B?
    b) Write down a similar statement for the velocity of B relative to A.

15 Two roads *ABC* and *DBE* meet at right angles. Peter is walking along *AB* at 6 km/h towards *B*. Paul is walking along *DB* at 6 km/h towards *B*. What is the velocity of

a) Paul relative to Peter
b) Peter relative to Paul?

The second diagram (not to scale) shows a vector solution to question *a*. Draw this diagram accurately and give the answer. Confirm your measurement by calculation. Draw a similar diagram for *b*.

c) What do you notice about the two relative velocities?

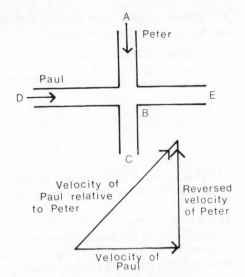

16 A ship *A* is sailing at 20 km/h on a bearing 090°. Another ship *B* is sailing at 15 km/h on a bearing 180°. What is the velocity of *B* relative to *A*? Either make an accurate scale drawing and measure your answer, or draw a rough sketch and calculate your answer using trigonometry.

17 A cyclist is travelling at 15 km/h along a straight road running due W. The wind is driving rain at a speed of 10 km/h from the NW.

a) At what angle does the rain appear to come (i.e. what is the direction of 'the velocity of the rain relative to the cyclist')?
b) Repeat *a* if the cyclist is moving at the same speed but due E.

18 In question 17, a sudden squall of wind brings the rain at twice its previous speed, but still from the NW. By how much has the wind apparently veered *a*) when the cyclist is travelling W   *b*) when he is travelling E at the speed stated?

19 The sketch shows a low-flying aircraft crossing a road at a speed of 180 km/h. A motorist is travelling along the same road at 90 km/h. The true path of the aircraft is *AC*, perpendicular to the line of the road. Does the aircraft appear to the motorist to be flying along a line such as *AB* or such as *AD*? Calculate the angle that this line makes with the line of the road, giving your answer to a tenth of a degree.

20 A motorboat crosses a straight stretch of river following a path which is perpendicular to the banks. The stream is running at 6 km/h and the speed of the boat relative to the water is 10 km/h. At what angle must it sail relative to the water in order to reach *B*? What is its true speed along *AB*? (*Hint* The next diagram (not to scale) shows the vector sketch for solving the problem.)

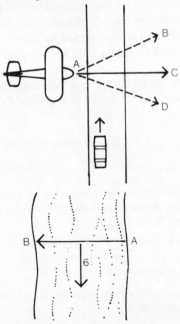

21 Repeat question 20 if the true path of the boat is a straight line making

a) 30° downstream with *AB*
b) 10° upstream with *AB*.

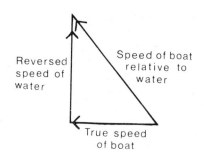

22 If in question 20 the boat is headed directly towards *B* (i.e. if the velocity relative to the water is along *AB*), at what angle with *AB* will its path actually be?

23 A large square is marked out on the deck of a ship. Its sides are parallel and perpendicular to the length of the ship. A man walks at 2 m/s forward across a diagonal of the square. The ship is sailing at 8 m/s. State *a*) the velocity of the man relative to the ship   *b*) his true velocity.

24 *A* and *B* are moving with velocities represented by the vectors stated. In each case find *i*) the velocity of *A* relative to *B*, *ii*) the velocity of *B* relative to *A*.

a) $2i+4j$, $i-j$ b) $\begin{pmatrix} 3 \\ 4 \end{pmatrix}$, $\begin{pmatrix} 2 \\ 7 \end{pmatrix}$ c) $\begin{pmatrix} -1 \\ 6 \end{pmatrix}$, $\begin{pmatrix} 3 \\ -5 \end{pmatrix}$

d) $3i-j$, $-2i+4j$ e) $4i$, $3j$

Combining Forces

25 Forces are usually combined using the 'parallelogram of forces'. If the forces are represented by the vectors $\overline{OA}$ and $\overline{OB}$ (*a* and *b*) then their combined effect (or 'resultant') is represented by the diagonal of the parallelogram *OACB*, i.e. by the vector *a+b*.

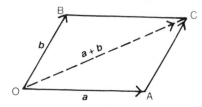

This can be shown just as clearly by the 'triangle of forces' as in the second diagram.

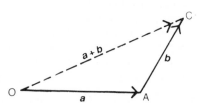

You are given the magnitude of each of two forces and the angle between their lines of action. Draw rough sketches to show *i*) the parallelogram of forces, *ii*) the triangle of forces for each pair.

a) 6 newtons, 10 newtons, 40°   b) 3 newtons, 4 newtons, 90°
c) 5 kg weight, 8 kg weight, 120°   d) 3 kg weight, 7 kg weight, 60°
e) 4 newtons, 6 newtons, 20°.

By accurate drawing or by calculation find the magnitude and direction of the resultant in *d*.

26 Two boys are dragging a loaded trolley along a road. Each is pulling on a separate rope attached to the front of the trolley, and each is exerting a force equal to the weight of 25 kg. If the ropes each make an angle of 10° with the direction of the road, draw a triangle of forces and find the effective pull on the trolley.

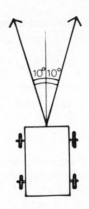

27 Two puppies, Shaggy and Scampy, and their mother Josie are sharing a marrow bone. Shaggy and Scampy are pulling with a force equal to ten times the weight of the bone, at an angle of 120° to each other. Draw a triangle of forces and find how hard Josie must pull to balance the other two.

28 A packing case weighing 30 kg rests on a sloping board which makes an angle of 30° with the horizontal. The weight of 30 kg acts vertically downwards as shown. The 'reaction' of the plane acts in a direction perpendicular to the plane and is equal to the weight of 26·0 kg. A rope holds the packing case from slipping down the board. If the force *P* in the rope is equal and opposite to the resultant of the other two forces, draw a triangle of forces and find the value of *P*.

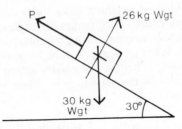

29 Anne and Susan are teaching John to roller-skate. He can keep upright but can only move slowly, so they take one arm each and tow him. If each pulls with a force of 10 newtons and their arms each make an angle of 25° with the direction of motion, with what total force is John being pulled?

*30 The three masses shown in the diagram are in equilibrium. Assuming that the force in each cord is equal to the weight of the mass hanging on the rope, find the value of *M*.

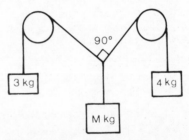

31 Pairs of forces represented by the vectors stated act on a body. Find their resultants.

a) $\binom{5}{1}$ and $\binom{4}{3}$ b) $\binom{2}{3}$ and $\binom{3}{1}$ c) $3i - 2j,\ 4j$

d) $6i,\ 2i - j$ e) $\binom{4}{0}$ and $\binom{0}{3}$

Work done by a Force

32 When a force P acts on a body and moves it through a distance d parallel to the line of action of the force, the work done is Pd. If the body moves at an angle θ to the line of action of the force, the work done is $P\cos\theta \times d$. If the force is represented by the vector p and the distance moved by the vector d then the work done is equal to $pd\cos\theta$, i.e. it is $p.d$.

a) If the force is 10 and the displacement 4, and the direction of the displacement makes an angle of $40°$ with the line of action of the force, what is the work done?

b) If the force is represented by the vector $\binom{2}{3}$ and the displacement by the vector $\binom{3}{4}$, what is the work done?

c) If the force is represented by the vector $3i + 5j$ and the displacement by the vector $4i + 2j$, what is the work done?

33 When a force of 1 newton acts on a mass and moves it through a distance of 1 m in the direction of its line of action, the work done is 1 joule. A solid mass weighing 10 kg slides down a smooth plank which makes an angle of $30°$ with the horizontal through a distance of 1 metre. What is the work done by gravity on the mass? (The force of gravity acting vertically on the mass may be taken as 100 newtons.)

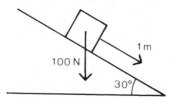

34 In each of the following, find the work done when a force represented by the first vector moves a body through a distance represented by the second vector.

a) $\binom{4}{3},\ \binom{2}{7}$ b) $3i + 2j,\ 2i$ c) $\binom{6}{2},\ \binom{8}{1}$

d) $3i + j,\ j$ e) $\binom{a}{b},\ \binom{c}{d}$

19 Abstract Algebra and the Algebra of Sets

19A Abstract Algebra

Note Some of the questions in 19A are revision questions.

	1	2	3	4	5	6
1	1	2	3	4	5	6
2	2	4	6	1	3	5
3	3	6	2	5	1	4
4	4	1	5	2	6	3
5	5	3	1	6	4	2
6	6	5	4	3	2	1

1 The table shows multiplication (mod 7) for the set $\{1, 2, 3, 4, 5, 6\}$. Answer the following questions:

a) Is multiplication commutative?
b) Is multiplication associative?
c) Name the identity element (i.e. the element e such that $ae = ea = a$ for every element a of the set $\{1, \ldots .6\}$).
d) State the inverse of 5 (i.e. find a number y such that $5 . y = y . 5 = e$).
e) Does every element of the set have an inverse?
f) Is the set of numbers $\{1, \ldots .6\}$ closed to multiplication (mod 7)?
g) Would your answers to questions a to f be true for any prime modulus (modifying the 'set of numbers' appropriately)?

2 Construct a similar table for multiplication (mod 6) for the set $\{1, 2, 3, 4, 5\}$. Answer questions a to f from question 1 for this non-prime modulus.

3 Construct a table for addition (mod 7) for the set $\{0, 1, \ldots .6\}$ and answer the following questions:

a) Is addition commutative? Is it associative?
b) Is there an identity element?
c) Does every element have an inverse?
d) Is the set of numbers $\{0, 1, \ldots .6\}$ closed to addition?
e) Using the set of numbers $\{0, 1, \ldots 5\}$ would your answers be true for addition (mod 6) (i.e. a non-prime modulus)?

4 Now consider the set of integers under ordinary multiplication.

a) Is there an identity element?
b) Does the element 4 have an inverse?
c) Does i) any element   ii) every element have an inverse?
d) Is the set closed to multiplication?
e) If the set was enlarged to include the rationals, would this affect any of your answers?

5 Consider the set of positive integers under subtraction.
a) Is subtraction of two elements always possible?
b) Is there an identity element?
c) Does any element have an inverse?
d) Is subtraction i) commutative   ii) associative for this set?
e) Is the set closed to subtraction?

6 If the set in question 5 were enlarged to include zero and the negative integers, would any of your answers to 5 a–d be affected?

7 If the set in question 5 remained as positive integers, but subtraction was replaced by $\sim$ (squiggles), where $a \sim b$ means 'take the smaller from the larger', would any of your answers to 5 a to d be affected?

8 *a)* What is the value of *i)* $4 \times (3+5)$ *ii)* $(4 \times 3)+(4 \times 5)$?
   b) Is multiplication distributive over addition for the integers?
   c) What is the value of *i)* $4+(3 \times 5)$ *ii)* $(4+3) \times (4+5)$?
   d) Is addition distributive over multiplication for the integers?
   e) Make up more examples of your own to illustrate your answers to *b* and *d*.

9 Copy and complete the multiplication table for these four matrices.

$$M_1 \begin{pmatrix} 0 & -1 \\ 1 & 0 \end{pmatrix} \quad M_2 \begin{pmatrix} 1 & 0 \\ 0 & 1 \end{pmatrix} \quad M_3 \begin{pmatrix} 0 & 1 \\ -1 & 0 \end{pmatrix} \quad M_4 \begin{pmatrix} -1 & 0 \\ 0 & -1 \end{pmatrix}$$

a) Name the identity element for this set under multiplication.
b) Give the inverse of M_1.
c) Does every element have an inverse?
d) Which elements are their own inverse?
e) Is multiplication *i)* commutative
                ii) associative for this set?
f) Is the set closed to multiplication?

	M_1	M_2	M_3	M_4
M_1	M_4	M_1		
M_2	M_1	M_2		
M_3	M_2			
M_4				

10 Repeat question 9 with these four matrices.

$$M_1 \begin{pmatrix} 1 & 0 \\ 0 & -1 \end{pmatrix} \quad M_2 \begin{pmatrix} -1 & 0 \\ 0 & -1 \end{pmatrix} \quad M_3 \begin{pmatrix} 1 & 0 \\ 0 & 1 \end{pmatrix} \quad M_4 \begin{pmatrix} -1 & 0 \\ 0 & 1 \end{pmatrix}$$

11 Consider the set of all 2×2 matrices.

a) Is multiplication always possible?
b) Is the set closed to multiplication?
c) Is the set closed to addition?
d) Is multiplication *i)* commutative *ii)* associative?
e) Is multiplication distributive over addition? Give an example.
f) Is addition distributive over multiplication? Give an example.

12 If *a* and *b* are integers and $a*b$ means 'square *a* and add *b*',
a) write down the values of $3*4$ and $4*3$.
b) is the operation $*$ commutative for the integers?
c) write down the value of *i)* $2*(3*2)$ *ii)* $(2*3)*2$.
d) is the operation $*$ associative for the integers?
e) is there an identity element, i.e. an element *e* such that $a*e=e*a=a$ for all *a*?
f) does *i)* any element *ii)* every element have an inverse?
g) are the integers closed to the operation $*$?

13 Repeat question 12 when $a*b$ means 'raise *a* to the power of *b*'.

14 Repeat question 12 when $a*b$ means 'add *a* to $3b$'.

15 Repeat question 12 when $a*b$ means 'divide *a* by *b*' and *a* and *b* are rationals.

16 A binary operation $*$ is defined on the real numbers by the formula $a*b=a+b-2$.

a) State whether the operation $*$ is commutative, giving reasons.
b) *i)* Calculate $3*4$ and $(3*4)*4$.
ii) Write down and evaluate another combination of the numbers 3 and 4 which, together with your answer to *i)*, will suggest whether or not the operation $*$ is associative.

c) Find the neutral (identity) element *e* such that $e*y=y=y*e$ for every possible value of *y*.

d) Find the inverse of the number 6, that is, the number *x* such that $x*6=e=6*x$, where *e* is the number found in *c* above. (MEI)

17 An operation * is defined on the *non-negative* real numbers by the formula $a*b=(a^2+b^2)$.

a) Calculate *i*) $3*4$ *ii*) $(3*4)*12$ *iii*) $3*(4*12)$.

b) What property do your results in *a* illustrate?

c) Find an element *e* such that for every element *x*, $e*x=x=x*e$.

d) Show that just one non-negative real number has an inverse with respect to this operation. (MEI)

18 The operation * is defined to mean 'the difference between' two numbers, ignoring negative signs. For example: $5*3=2$; $1*4=3$. Working entirely within the set $S=\{0, 1, 2, 3, 4\}$ answer the following questions:

a) Evaluate *i*) $1*3$ *ii*) $(1*3)*4$ *iii*) $1*(3*4)$.
State with reasons whether or not the operation * is associative.

b) State the identity (neutral) element of *S* under the operation, giving two examples to justify your statement.

c) For an element *p* of *S*, state with reasons the inverse of *p* under the operation *. (MEI)

19B   The Algebra of Sets

The Basic Laws

Law 1: The Commutative Law
$A\cup B=B\cup A$
$A\cap B=B\cap A$

Law 2: The Associative Law
$A\cup(B\cup C)=(A\cup B)\cup C=A\cup B\cup C$
$A\cap(B\cap C)=(A\cap B)\cap C=A\cap B\cap C$

Law 3: The Distributive Law
$A\cup(B\cap C)=(A\cup B)\cap(A\cup C)$
$A\cap(B\cup C)=(A\cap B)\cup(A\cap C)$

Law 4: The Equal Power Law
$A\cup A=A$
$A\cap A=A$

Law 5: The Inclusion Law
$A\cup(A\cap B)=A$
$A\cap(A\cup B)=A$

Law 6: The Laws for ϕ and $\mathscr{E}$
$A\cup\phi=A$ $A\cap\mathscr{E}=A$
$A\cap\phi=\phi$ $A\cup\mathscr{E}=\mathscr{E}$

Law 7: The Laws for Complements
$A\cup A'=\mathscr{E}$ $A\cap A'=\phi$ $(A')'=A$
$\phi'=\mathscr{E}$ $\mathscr{E}'=\phi$

Law 8: The de Morgan Laws
$(A\cup B)'=A'\cap B'$
$(A\cap B)'=A'\cup B'$

1   In laws 1, 2, 3, 4, 5 and 8, how could you obtain the second law of each pair from the first? These pairs of laws are said to be 'duals'.

2   Can you see duals in laws 6 and 7? How would you obtain one member of the pair from the other in laws 6 and 7?

3   Using laws 1 to 8, the truth (or otherwise) of statements about sets can be deduced.

Example   To show that $A\cap(A'\cup B)=A\cap B$,

$$A\cap(A'\cup B)=(A\cap A')\cup(A\cap B) \qquad \text{(Law 3)}$$
$$=\phi\cup(A\cap B) \qquad \text{(Law 7)}$$
$$=A\cap B \qquad \text{(Laws 1 and 6)}$$

165

Prove similarly: *a)* $(A \cup B')' = A' \cap B$ *b)* $A' \cap (A \cup B') = A' \cap B'$
c) $B \cup (A \cap B') = B \cup A$ *d)* $(A \cap B') \cup A' = (A \cap B)'$

4 Prove the following results. Confirm each by using a Venn diagram or a truth table.

a) $A' \cup B' \cup (A \cap B) = \mathscr{E}$ *b)* $(A \cap B) \cap (A \cup B) = A \cap B$
c) $(A \cap B) \cup (A \cap C) = A \cap (B \cup C)$ *d)* $(A \cup B) \cap A = (A \cap C) \cup A$
e) $(A \cup B \cup C)' = A' \cap B' \cap C'$ *f)* $(A' \cap B) \cup (A \cap B) = B$

5 Prove the truth or otherwise of the following statements, not all of which are true.

a) $(B' \cup C') \cup (B \cup C) = \mathscr{E}$ *b)* $(A \cup B) \cap (A' \cap C)' = A \cup (B \cap C)$
c) $(A \cup B) \cap C = A \cup (B \cap C)$ *d)* $(A' \cap B') \cap B = \phi$
e) $[(A' \cup B') \cap C] \cup (A \cup B) = A \cup B \cup C$
f) $(A' \cap B' \cap C') \cup (A \cup B \cup C) = \mathscr{E}$
g) $(A \cup B) \cap (A \cup C) \cap (B \cup C) = (A \cap B) \cup (A \cap C) \cup (B \cap C)$
h) $C' \cap (A \cup B)' = A' \cap B' \cap C$

Confirm your findings in the first four by using truth tables and in the last four by using Venn diagrams.

6 Using set symbols, give an expression for each of the following regions in the Venn diagram shown:

a) 8   *b)* $3+5+7$ *c)* $2+4$ *d)* $2+3+5$

Example   Region 1 is $(A \cup B \cup C)'$.

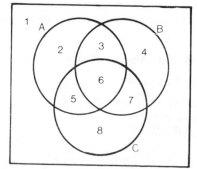

7 Assuming the truth of some of these laws, it is possible to prove the other laws from them.
Assuming the truth of laws 1, 2, 3, 4, 6 and 7, prove the truth of *a)* Law 5
b) Law 8.
(*Hint for Law 8*   Prove that $(A \cup B) \cup (A' \cap B') = \mathscr{E}$ and that $(A \cup B) \cap (A' \cap B') = \phi$. It follows that $(A \cup B)$ is the complement of $(A' \cap B')$.)

     Example   Start the proof for law 5 like this:
$$A \cup (A \cap B) = (A \cap \mathscr{E}) \cup (A \cap B) \qquad \text{(Law 6)}$$
$$= A \cap (\mathscr{E} \cup B) \qquad \text{(Law 3, reversed)}$$

8 In the algebra of sets, state the identity element for *a)* union, *b)* intersection.

9 In the algebra of sets, is there an inverse for set *A* under *a)* union, *b)* intersection? Give reasons for your answers.

10 *a)* In ordinary algebra, multiplication is distributive over addition, but addition is not distributive over multiplication. Give examples.
     b) The algebra of sets is 'doubly distributive'. Union is distributive over intersection and intersection over union. State this in symbols for the sets *A* and *B* and *C*. Show it also for the sets $A = \{3, 4, 5, 6, 7\}$, $B = \{2, 4, 6\}$ and $C = \{3, 6, 8\}$.

20 Boolean Algebra and Switching Circuits

Introduction

In 1854 George Boole published *The Laws of Thought* in which he showed that many of the basic ideas of logic could be represented by symbols and these could be manipulated by an 'algebra' whose laws were similar to (but not identical with) those of ordinary algebra. Before this time algebra had been applied almost exclusively to numbers, so Boolean algebra heralded the start of a new era in mathematics.

In 20A we show how logical statements can be expressed as symbols. In 20B we give the laws of Boolean algebra and examples of their manipulation. 20C deals with switching circuits.

20A Preparing the Ground

1 Statements

A statement is a declaration that can be either true or false, but not both. 'All bottles are blue' and 'All cats have five legs' are both statements, although both happen to be false. 'Green grass' and 'Do you like bananas?' are not statements. Statements are also known as *propositions*. Which of the following are statements?

a) My car has broken down. b) Cats catch mice. c) The house that Jack built.
d) Were you late for school this morning? e) Some donkeys have two legs. f) Blue skies. g) Little lambs eat ivy.

2 Compound Statements

Here are eight statements each of which is represented by a letter.

All cows can fly. (p)   I like apples. (q)   All apples have pips. (r)   All citrus fruits have pips. (s)   I do not like pips. (t)   Some horses like apples. (u)   Every cat has seventeen legs. (v)   Every animal with seventeen legs is a cat. (w)

The statement 'I like apples *and* all apples have pips' is a *compound statement* and can be represented by $q \cap r$.

The statement 'Either I like apples *or* all apples have pips *or* both' is another compound statement and can be represented by $q \cup r$.

Express the following compound statements (or compound propositions) in symbols:

a) Every cat has seventeen legs and all citrus fruits have pips.
b) Either every cat has seventeen legs or all cows can fly or both.
c) Either I like apples or I do not like pips or both.
d) I like apples and I do not like pips.
e) Some horses like apples and all cows can fly.

Write in words   f) $v \cup s$ g) $r \cap s$ h) $q \cap t$ i) $p \cup w$.

3 Negating Statements

a) The negation of a statement a is a', the complement of a. It can also be expressed in words. Thus the negation of statement a in question 2 is 'Not all cows can fly' or 'Some cows cannot fly'. Would it be correct to say that a' is 'No cows can fly'?

b) Using the statements in question 2, write in words:

i) q' ii) r' iii) v' iv) w'

c) Write the following in symbols:

i) Some citrus fruits do not have pips.

ii) Some cats do not have seventeen legs.

4

Using the propositions from question 2, express the following compound propositions in words:

a) $p \cup q'$ b) $r' \cap s$ c) $t' \cap q'$ d) $u \cup v'$ e) $r \cap v'$

5 Implication

$a \Rightarrow b$ means 'a implies b' or 'If a is true, then b is true also' or more simply 'If a, then b'.

$a \Leftrightarrow b$ means 'If a is true, b is also true and if b is true, then a is also true'.

Using the same statements as in question 2, express in words:

a) $s' \Rightarrow w$ b) $t \Rightarrow u'$ c) $p' \Rightarrow v$ d) $u \Leftrightarrow r$ e) $q' \Leftarrow r$

Additional examples of the use of $\Rightarrow$ and $\Leftrightarrow$ are given in nos. 10–14.

6 Compound Propositions and Truth Tables

A compound proposition can be represented by a truth table. If there are two statements, the table will contain four lines, if three statements, eight lines, etc. 1 is used for a true statement and 0 for a false statement. The truth table opposite shows the compound proposition $a \cap b$. Write down truth tables for the compound propositions

a	b	$a \cap b$
1	1	1
1	0	0
0	1	0
0	0	0

a) $c \cup d$ b) $e \cap f'$ c) $g' \cup h'$.

7 Equating Statements

Two statements are said to be equal if they are logically equivalent. Thus the statements 'All girls have long hair and all boys ride bicycles' is equal to the statement 'All boys ride bicycles and all girls have long hair' but it is not equal to the statement 'All boys and girls have long hair and ride bicycles'.

Are any of the following statements equal?

a) Cows eat grass and sheep eat grass.   b) Both cows and sheep eat grass.   c) Boys over 16 can leave school.   d) If you are a boy, you can leave school when you are over 16.   e) If you can leave school, you are a boy over 16.   g) A new motorcar costs more than £1000.   h) If a motorcar costs more than £1000, it is a new one.

8 Investigating the Equality of Statements using Truth Tables

The truth table for the statement $p' \cup q$ is shown opposite. 1 indicates that a statement is true and 0 that it is false. A truth table for two statements contains four lines. How many lines will there be for three statements?

p	q	p'	$p' \cup q$
1	1	0	1
1	0	0	0
0	1	1	1
0	0	1	1

The truth table for $p \Rightarrow q$ is also shown. To understand it, consider the statement made by a father to his son: 'If you polish my shoes I'll give you 10p.' Let $p =$ 'you polish my shoes' and $q =$ 'I'll give you 10p.' In the first line the son polishes the shoes and gets 10p. The promise is true. In the second line, though he polishes the shoes he doesn't get 10p, so the promise is false. In the third and fourth lines, the shoes have not been polished but the promise has not been broken and is therefore taken as true. As the last column of this table and the one above are equal, the two statements are equal, i.e. $p \Rightarrow q$ is equal to $p' \cup q$.

p	q	$p \Rightarrow q$
1	1	1
1	0	0
0	1	1
0	0	1

Now use truth tables to show that

a) $(p \Leftarrow q) = p \cup q'$ b) $(p \Leftrightarrow q) = (p' \cup q) \cap (p \cup q')$.

Further examples of truth tables being used to examine the truth of compound propositions are given in 20B.

9 Tautologies

A compound statement that is always true (whatever the truth values of its individual statements) is called a *tautology*. Thus the statement 'I have got an apple or I have not got an apple' is a tautology. $a \cup a'$ is always a tautology and this is written $a \cup a' = 1$.

a) What will be the last column of the truth table of a tautology?
b) Which of the following are tautologies:
i) Either all horses like apples or some horses do not like apples.
ii) This parcel weighs more than one kilogram or it weighs less than one kilogram.
iii) $e \cup (e' \cup f)$ iv) $(a \cap b) \cup (a \cap b)'$

Further Examples on the Use of $\Rightarrow$ and $\Leftrightarrow$

10 Which of the following are true and which are false?

a) $(x = 5) \Rightarrow (x < 6)$ b) $(x + 4 = 7) \Rightarrow (x = 3)$
c) $(x = 2) \Rightarrow (x^2 - 5x + 6 = 0)$ d) $(x = 2) \Leftrightarrow (x^2 - 5x + 6 = 0)$
e) For the quadrilateral $ABCD$, $ABCD$ is a square$\Leftrightarrow$angle A is 90°.

11 If $x \in \mathscr{E}$ and A and B are subsets of $\mathscr{E}$, state whether the following are true or false:

a) $x \notin A \Rightarrow x \in A'$ b) $A \subset B \Leftrightarrow B \supset A$ c) $A \cap B = \phi \Leftrightarrow A \not\subset B$
d) $A \cap B = A \Leftrightarrow A \subset B$
e) All children like a game and all dogs like a game $\Rightarrow$ children are dogs.

12 Insert $\Rightarrow$ or $\Leftarrow$ or $\Leftrightarrow$ between these pairs of statements to make the combined statements true.

a) $x + 4 = 7$, $x = 3$ b) $x^2 = 16$, $x = 4$ c) $A \cup B = A \cap B$, $A = B$
d) Angles B and C of triangle ABC are acute, angle A is a right angle.

13 Insert $\Rightarrow$ or $\Leftarrow$ or $\Leftrightarrow$ between these pairs of statements to make the combined statements false:

a) $x = 5$, $x > 3$ b) $x^2 = 16$, $x = 4$ c) $x \notin A'$, $x \in (A \cup B)$
d) A tall student had a brother Tom, Tom had a brother who was a tall student.

14 The sets P, Q and R, which are subsets of the universal set, are related as shown in the diagram. The following are six statements concerning these sets:

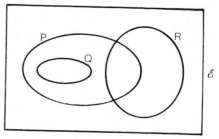

i) $x \in Q' \Rightarrow x \in P'$ iv) $x \in P \Rightarrow x \in Q$
ii) $x \in R \Rightarrow x \in Q'$ v) $x \in P \cap R \Leftrightarrow x \in Q'$
iii) $x \in Q \Rightarrow x \in P$ vi) $x \in Q \Rightarrow x \in R'$

a) State which of the above statements are true and which are false.

b) There are two pairs of logically equivalent statements in the above six. Which are they?

(MEI)

20B The Laws of Boolean Algebra

We are now ready to use the laws of Boolean algebra. Several notations are in use. Thus $a \cap b$ can be written $a.b$ or even ab, and $a \cup b$ can be written $a+b$. 1 indicates a tautology. 0 indicates a statement that is false whatever the truth values of its individual components, i.e. the last column of its truth table is all zeros. For the rest of this exercise we shall use . and $+$ instead of $\cap$ and $\cup$.

1   The Commutative Laws
$a+b=b+a$
$a.b=b.a$

2   The Associative Laws
$a+(b+c)=(a+b)+c=a+b+c$
$a.(b.c)=(a.b).c=a.b.c$

3   The Distributive Laws
$a+b.c=(a+b).(a+c)$
$a.(b+c)=a.b+a.c$

4   The Equal Power Laws
$a+a=a$
$a.a=a$

5   The Inclusion Laws
$a+a.b=a$
$a.(a+b)=a$

6   The Laws for 1 and 0
$a+0=a$ $a.1=a$
$a+1=1$ $a.0=0$

7   The Laws for Complements
$a+a'=1$ $a.a'=0$ $(a')'=a$
$0'=1$ $1'=0$

8   The de Morgan Laws
$(a+b)'=a'.b'$
$(a.b)'=a'+b'$

1   Look at these laws and answer the following questions:

a) Are . and $+$ doubly distributive?
b) Which part of which law illustrates a tautology?
c) Which laws are exactly similar to the laws for $+$ and $\times$ in ordinary algebra?
d) Name three laws which would be incorrect in ordinary algebra.
e) Name two laws which have no counterpart in ordinary algebra.

2   The laws of Boolean algebra can be used to simplify compound statements or to find out whether two compound statements are equal.

Examples   Simplify $a.b+a.b'+a$

$$a.b+a.b'+a=a.(b+b')+a \qquad \text{(Law 3)}$$
$$=a.1+a \text{ (Law 7)} \quad =a+a \text{ (Law 6)} \quad =a \text{ (Law 4)}$$

Prove that $(b+c)'+a'=(a'+b').(a'+c')$
$$(b+c)'+a'=(b'.c')+a' \qquad \text{(Law 3)}$$
$$=a'+(b'.c') \qquad \text{(Law 1)}$$
$$=(a'+b').(a'+c') \qquad \text{(Law 3)}$$

Simplify a) $a.b + b.a' + (b+c)$ b) $a'.b + a.b$

c) $b' + (a'+b)$ d) $a.b.(a'+b')$

Prove e) $(a'+b)'.a = a.b'$ f) $(a+b).(a+c) = a+b.c$

g) $a.b + a = (a+c).a$ h) $a.b'.c' + b.a + c.a = a$

i) $b'.c' + (b+c) = 1$

3 Using Boolean algebra, find out whether the following are tautologies. Check your answers using truth tables. (In the case where the proposition is not a tautology, it is only necessary to find one line of the truth table which gives a zero in the last column.)

a) $(a+b) + a'.b'$ b) $(b+c) + a.b'.c' + a'.b'.c'$

c) $(c+a) + b.a'.c' + a'.b'.c'$ ✱ d) $b+c+a.b'.c'$

✱ e) $(a+c') + b.(a+c)' + a'.b'.c$

4 Use Boolean algebra to prove the truth of the following statements. Confirm by using truth tables.

a) $b.c.(b+c) = b.c$ b) $(a+b+c)' = a'b'c'$

c) $a.b.c + a.b'.c = (a'+c')'$ d) $(b+c).(b'+c') \Rightarrow b+c$

e) $a'.b.c' + a'.b'.c' = (a+c)'$ f) $(a+b)'.c + a'.b'.c' = a'.b'$

g) $(a+b') + (a'+b')' = b$ h) $a.b'.c + a.b.c \Rightarrow a.c$

i) $a.b + a'.b.c' + a'.b.c = b$ j) $a'.b.c' + a.b.c' + a.b.c + a'.b.c = b$

5 Venn diagrams form a useful check on Boolean algebra. In the diagrams some areas are marked. Copy them and mark the remaining areas similarly.

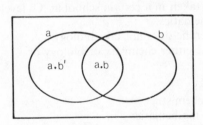
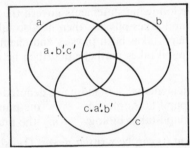

6 A mother uses the following ingredients (among others) when making cakes: baking powder (b), cherries (c), dates (d), golden syrup (g), mixed fruit (m), plain flour (p), self-raising flour (r), sugar (s).

If in a particular recipe she uses plain flour, this is denoted by p.

If she does not use cherries when she uses dates, this is denoted by $d \Rightarrow c'$.

Express in words the meanings of the following statements:

a) $m+c$ b) mc c) $r \Rightarrow b'$ d) $g' \Rightarrow s$ e) $p+r$ f) $r \cup p$ g) $(c' \cup m) \cap (m' \cup c)$.

Express in symbols the following statements:

h) Sometimes she uses golden syrup, sometimes sugar, and sometimes both.

i) When she uses plain flour, she uses baking powder.

j) She does not use dates without also using golden syrup.

State in both words and symbols the converse of statements c and d above.

7 In a certain school there are a number of after-school clubs. These include chess (c), football (f), gymnastics (g), hockey (h), cookery (k), needlework (n) and swimming (s). If a pupil joins the chess club, this is denoted by c. If the pupil does not join, it is denoted by c'. Express in words the meaning of the following statements:

a) $c.s \Rightarrow g'$ b) $h \Rightarrow f'$ and $f \Rightarrow h'$ c) $n+c \Rightarrow k'$ d) $s+f+c$ e) $s.f.h$ f) $f \Rightarrow g$

8 In this question *u* and *v* are positive integers and *p* and *q* are the following statements: *p*:*u* is an even number; *q*:*v* is an even number.
The condition for *uv* to be even is that at least one of *u* and *v* must be even. We write this condition as *p+q*.

a) Write down the condition for *i)* *u* to be an odd number *ii)* *uv* to be an odd number.
b) The condition for *u+v* to be odd can be written *pq′+p′q*. Express this condition in words.
c) Find an alternative form for the condition for *u+v* to be odd by finding the condition for *u+v* to be even and negating it.
d) Construct truth tables for the condition in *b* and for your answer to *c* and hence show that they are equivalent.

(MEI)

9 The letters *a* and *b* represent a single digit, 0 or 1. The letter *d* represents the difference between *a* and *b*, and the letter *s* represents the sign of (*a−b*) in the following way: when (*a−b*) is negative, *s* takes the value 1; when (*a−b*) is positive or zero, *s* takes the value 0.

a) Copy and complete the table shown.
b) Write down Boolean formulae for *s* and *d* using some or all of the symbols *a*, *b*, *a′·b′*.
c) Reduce the Boolean expression (*a+b*)(*a′+b*) to its simplest form.   (MEI)

a	*b*	(*a−b*)	*s*	*d*
0	0	(0)	0	0
0	1	(−1)	1	1
1	0			
1	1			

10 The following statements govern the subjects taken in a certain school at 'O' level:
i) If a pupil takes physics, then he also takes mathematics.
ii) If a pupil takes both physics and chemistry, then he does not take biology.
iii) If a pupil takes mathematics, then he must take either chemistry or biology or physics.

The statements *m*, *p*, *b* and *c* are denoted as follows:
p   the pupil takes physics   *c*   the pupil takes chemistry
b   the pupil takes biology   *m*   the pupil takes mathematics

Statement (*i*) can be written *p⇒m*. Translate statements (*ii*) and (*iii*) into symbols. State both in ordinary English and in symbols the converse of statement (*i*).   (MEI)

20C   Switching Circuits

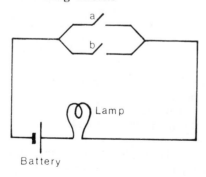
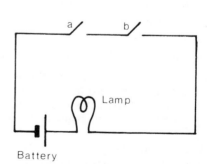

1   The first diagram shows a circuit made up of a battery, a lamp and two switches *a* and *b* in parallel. The second diagram shows a similar circuit with the switches *a* and *b* in series.

a) In the first circuit will a current pass if *i*) *a* only is closed   *ii*) *b* only is closed
iii) *a* and *b* are both closed   *iv*) neither *a* nor *b* is closed?
b) Answer the same questions as in *a* for the second circuit.

2   There is a striking similarity between switching circuits and the union and intersection of sets. Consider the two tables below:

a	*b*	*a* and *b* in parallel	*a* and *b* in series
1	1	1	1
1	0	1	0
0	1	1	0
0	0	0	0

A	*B*	$A \cup B$	$A \cap B$
1	1	1	1
1	0	1	0
0	1	1	0
0	0	0	0

In the first table, 1 in either of the first two columns means that the switch is closed and 0 means that the switch is open. In the third and fourth columns, 1 means a current is flowing and 0 means it is not.

The second table is of the type we have already studied in chapter 2.
Now answer the following questions:

a) Do switches in parallel give a similar arrangement of 1's and 0's to $A \cap B$ or to $A \cup B$?
b) Would $a \cap b$ represent switches *a* and *b* in series or in parallel?
c) What would $a \cup b$ represent?

3   It is more usual to represent switching circuits in Boolean algebra than in the algebra of sets. So *a.b* would represent switches *a* and *b* in series, and *a+b* switches *a* and *b* in parallel. State what the following Boolean expressions would represent:

a) $p.q.r$ *b*) $w+x+y$ *c*) $(p+q).r$ *d*) $w.(x+y)$
Draw diagrams to represent each of your answers.

4   Write down Boolean expressions to represent the following circuits.
The battery and lamp are no longer included in the diagrams.

a)
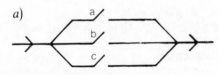

b)
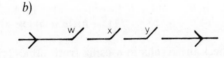

c)
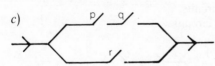

d)
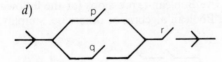

e)
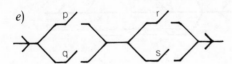

f)
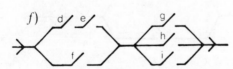

5   Draw circuits to represent the following Boolean expressions:

a) $a.b+(c+d)$ *b*) $w.x.y+z$ *c*) $(r.s+t.u).v$

6 If two switches are linked so that both are closed together or both open together, they are represented by the same letter in a switching circuit diagram or in a Boolean expression. If they are *contra-linked* so that when one is closed the other is open and vice versa, the same letter is used for both but a dash is added to one. In the following diagram, pick out pairs of switches that are directly linked and pairs that are *contra-linked*:

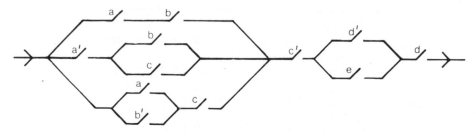

7 Using the laws of Boolean algebra, simplify the following circuits. In each case draw the original circuit and the final simplified circuit.

   a) $p.p'$ *b)* $p+p'$ *c)* $(p.p')+p+p'$ *d)* $(p+p').p'$

8 A switching circuit is represented by the Boolean expression $a.b+a.b'$. The expression can be simplified as follows:

$$\begin{aligned} a.b+a.b' &= a.(b+b') & \text{(Law 3)} \\ &= a.(1) & \text{(Law 7)} \\ &= a & \text{(Law 6)} \end{aligned}$$

Original circuit   *Simplified circuit*

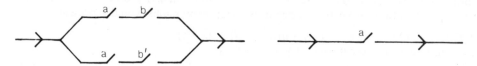

Simplify the following circuits, giving two diagrams for each:

   a) $a+a'.b$ *b)* $a.b.(a+b)$ *c)* $c'+a+c.a'$ *d)* $a.b'+a'.b'$
   e) $a.b+a'+a'.b'$ *f)* $(a+b).(a'+b)$ ✱ *g)* $a.c+b.c+a'.b'$ ✱ *h)* $a.b'+b.c'+c.a$

9 Check the results in **8** using truth tables (see question **2**).

10 Give Boolean expressions for the following circuits.
Using Boolean algebra, or otherwise, simplify the circuits.

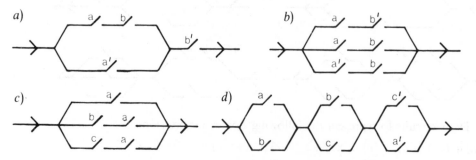

174

e) 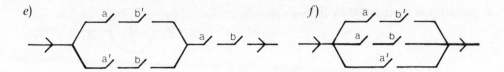 *f)*

11 Check your answers to question 10 using truth tables.

12 Draw the switching circuit represented by the Boolean function $(A+B).(A'+B')+A'.B'$. By simplifying the Boolean function, or otherwise, find the simplest equivalent circuit.

13 Draw the switching circuit corresponding to the Boolean function $a.(b+c)+b.(c+a')+a.c'+c.b'$. By simplifying this Boolean expression, or otherwise, find the simplest equivalent circuit.

14 In the given circuit the Boolean condition for the light B to be on is pq. What is the condition for the light A to be on?

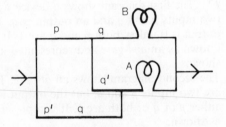

Copy and complete this table which refers to the given circuit where 0 indicates that a switch is open (and a light off) and 1 indicates the reverse condition. (MEI)

p	q	B	A
0	0		
0	1		
1	0		
1	1	1	0

15 Copy and complete the given truth table. Write down a formula in Boolean algebra which can be deduced from the table.
Write down a Boolean expression to represent the given circuit. Simplify the expression as far as possible and draw the resulting simplified circuit. (MEI)

p	q	$p+q$	p'	$p'q$	$p+p'q$
0	0	0	1	0	0
0	1	1	1	1	
1	0				
1	1				

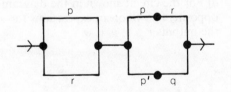

16 *Invertors and Gates*
A device which will change an input x into an output x' is known as an invertor. Copy and complete the table opposite.

Input x	Output x'
1	
0	

An invertor is represented by the sign shown:

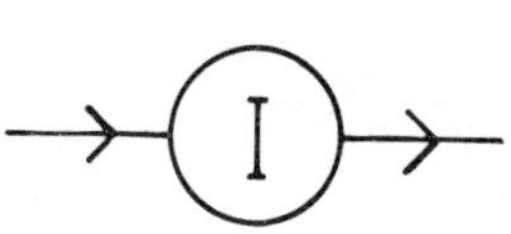

One form of invertor is shown in the diagram. A magnetically controlled switch is operated by the input current, and breaks the circuit for the output which has an independent source of current.

OUTPUT x'

INPUT

x

switch

Independent source of current for output

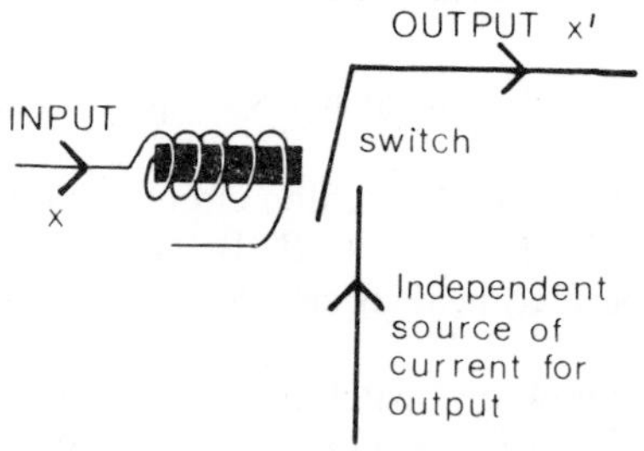

17 The first diagram shows a device with two inputs p and q and an output $p.q$. This output is 0 unless both p and q are 1. It is known as an *and-gate* and represented as shown.

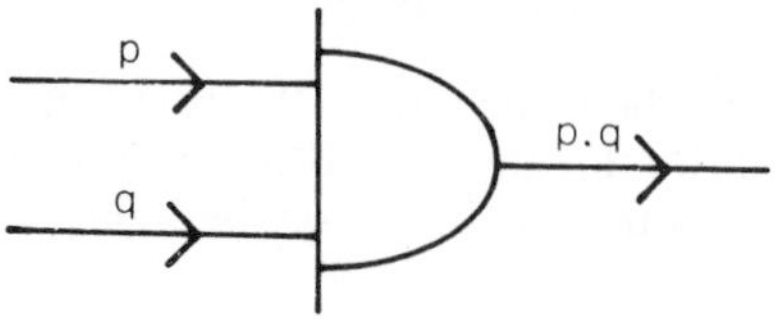

The second diagram shows an *or-gate*. There are two inputs p and q, but the output is 1 if either p or q or both are 1. It is represented as shown.

a) What is the correct Boolean expression for the output of the *or-gate*?

b) Copy and complete the table opposite.

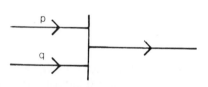

p

q

p.q

p

q

Input		Output	
p	q	and-gate	or-gate
1	1		
1	0		
0	1		
0	0		

18 *a)* For the circuit shown in the diagram opposite, find Boolean expressions for the outputs r, s, t, u, v, w.

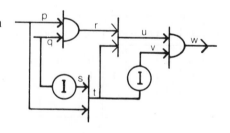

p
q
r
u
v
w
s
t
I
I

b) Copy and complete the table. Does the last column agree with the expression you found for w?

p	q	r	s	t	u	v	w
1	1	1	0				
1	0						
0	1						
0	0						

19 Repeat question 18 for the circuit shown, using the letters $p, q, \ldots v$.

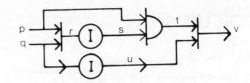

20 A NAND circuit element has the following properties:

i) If there is one input p it gives the single output p'.

ii) If there are two inputs p and q, it gives the single output $(pq)'$.

a) Copy and complete this table for the output from a single NAND element with inputs p and q.

b) The following circuit has inputs p and q, two intermediate outputs r and s, and a final output t. Copy and complete the table. What is the usual way of writing the output in terms of p and q?

p	q	Output
1	1	
1	0	
0	1	
0	0	

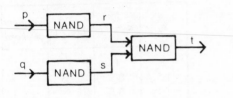

p	q	r	s	t
1	1			
1	0			
0	1			
0	0			

c) Copy and complete the table for the following circuit:

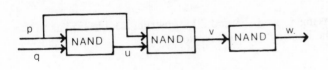

p	q	u	v	w
1	1			
1	0			
0	1			
0	0			

(MEI)

21 The subjects mathematics, physics, chemistry and biology may be taken at 'O' level in various combinations by pupils in a certain school. Design a switching circuit that will indicate if a pupil takes both mathematics and at least one science subject. Your circuit should consist of four switches labelled m (for mathematics), p, c and b, and also a lamp and battery. If a pupil takes any of these subjects, he depresses the appropriate switches. The lamp should light if he takes both mathematics and at least one science subject.

(MEI)

21 Calculus

21A The Slope of a Tangent

1 The diagram shows the points A (1, 1) and B (2, 4) on the curve $y=x^2$.

a) Write down the gradient of the chord AB.

b) C is a third point on the same curve, with x co-ordinate 1·5. What is its y co-ordinate? Write down the slope of AC.

c) D is a fourth point on the same curve with x co-ordinate 1·1. Write down the gradient of AD.

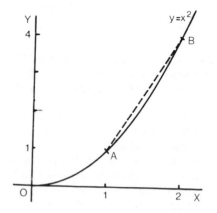

2 a) Using the results of question 1, copy and complete the table shown. (x_1, y_1) are the co-ordinates of A and (x_2, y_2) the co-ordinates of B, C or D.

b) (*Optional*) Add two more lines to the table taking x_2 as 1·01 and 1·001 respectively.

x_2	$x_2 - x_1$	$y_2 - y_1$	Gradient of chord
2	1	3	3
1·5	0·5	1·25	
1·1	0·1		

c) The gradient of the chord gets less as $x_2 - x_1$ gets smaller. But the gradient does not decrease indefinitely. It 'approaches a limit' as $x_2 - x_1$ approaches 0. What is this limit?

3 Starting with $x_1 = 2$ and taking x_2 as 3, 2·5 and 2·1, construct a table similar to the table in question 2. What is the limiting value of the slope?

4 Repeat question 3 starting with $x_1 = 3$ and $x_2 = 4$, 3·5 and 3·1.

5 Comparing the results of questions 2, 3 and 4, can you see a simple relation between the limiting value of the gradient and the value of x_1?

6 The diagram shows the chords AB, AC in question 1 replaced by *secants*, i.e. by chords extended in both directions. As the chords get shorter, what do the secants become 'in the limit'? What does the limiting value of the slopes in questions 2, 3 and 4 represent?

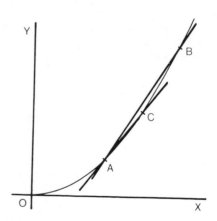

7 The diagram shows P and Q, two adjacent points on the curve $y=x^2$. P is (x, y) and q is $(x+a, y+b)$.

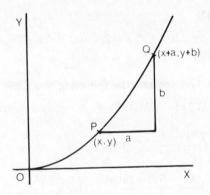

a) Write down a relationship between x, y, a and b and, remembering that $y=x^2$, get an expression for b in terms of x and a.
b) The gradient of the chord PQ is b/a. Express this in terms of a and x.
c) Remembering that a is the same as x_2-x_1, does this gradient agree with the gradients in the last columns of the tables in questions 2, 3 and 4?

8 Taking A as the point whose x co-ordinate is 1 and B as the point whose x co-ordinate is 2, repeat question 1 for the following curves:

a) $y=x^2+3x$ b) $y=2x^2+3x$ c) $y=x^2+2x-4$

9 Repeat question 7 for each of the curves in question 8, recording as your answers the values of the gradient b/a. Do your answers agree with the gradients found in question 8?

10 P and Q are adjacent points on the curve $y=x^2$. The co-ordinates of P are (x, y) and of Q $(x+\delta x, y+\delta y)$ where δx and δy (pronounced 'delta x' and 'delta y') are small increments in x and y.

a) Find the gradient $\delta y/\delta x$ for the chord PQ. Give your answer in terms of x and δx.
b) If $\delta x \to 0$ (read this as 'if the value of δx approaches zero'), the gradient of the chord becomes the gradient of the tangent at P, i.e. the gradient of the tangent at P is the limiting value of $\delta y/\delta x$. This is written dy/dx (read it as 'dy by dx'). Write down the value of dy/dx, the gradient of $y=x^2$, at the point P.

Note δy, δx, $\delta y/\delta x$ and dy/dx are all single quantities. The first two represent single lengths, and the second two represent single slopes.

∗ 11 Using the method of question 10, find the gradients of each of the following curves:

a) $y=2x^2$ b) $y=3x^2$ c) $y=x^2+x$ d) $y=x^2+2x$ e) $y=x^2+4$
f) $y=x^2-2$ g) $y=1-x^2$ h) $y=2-3x-x^2$

∗12 Using the method of question 10, find dy/dx for $y=x^3$.

Hint $(x+a)^3$ is $x^3+3ax^2+3a^2x+a^3$.

∗13 Repeat question 12 for $y=x^4$.

14 If $y=x^n$, $dy/dx=nx^{n-1}$. (The proof of this is beyond the scope of this book.) Using this formula, write down dy/dx for $y=x^2$, $y=x^3$, $y=x^4$.
Do these values agree with the values found in questions 10, 12 and 13?

15 Using question 14, if $y=1/x$ find dy/dx. (*Hint* Write $1/x$ as x^{-1}.)

16 If $y=1/x^2$, find dy/dx.

21B

Note If $y=x^n$, $dy/dx=nx^{n-1}$ for all values of n, positive or negative, integral or fractional.

1 Differentiate the following functions with respect to x:

a) $3x^2$ b) $3x^2+2x$ c) $\dfrac{x^2}{2}$ d) $\dfrac{1}{x^2}$ e) $5-3x-4x^2$

f) $\dfrac{1}{x^3}-\dfrac{1}{x^2}$ g) $x(4x-1)$ h) $\dfrac{5}{x^4}$ i) $\dfrac{2}{x}-\dfrac{3}{x^2}+\dfrac{4}{x^3}$

j) $\dfrac{3x^3-5x^2+2x}{x}$ k) $\left(x+\dfrac{1}{x}\right)^2$ l) $(5-x)(2+x)$

2 If $y=4x^3-2x^2+1$, find dy/dx. By substituting $x=-1$ into the expression for dy/dx, find the gradient of the curve at the point where $x=-1$.

3 Find the two points at which the curve $y=(x-2)(x-3)$ cuts the x axis. Find also the gradient of the curve at each of these two points.

4 Repeat question 3 for the following curves:

a) $y=(2x-1)(x+3)$ b) $y=x^2+3x-10$ c) $y=\dfrac{1}{x}-x$

d) $y=x^3-4x^2+3x$ (3 points) e) $y=4-\dfrac{1}{x^2}$

5 For each of the following curves, find the x co-ordinate of the point where $dy/dx=0$.

a) $y=x^2-x-6$ b) $y=(x+4)(x-2)$ c) $y=x^2+3x$
d) $y=4+3x-x^2$ e) $y=2x^2-3x-1$

What can you say about the tangent to a curve at the point where $dy/dx=0$?

6 Find the co-ordinates of the point on the curve $y=x^2+5x+6$ at which $dy/dx=1$. What angle does the tangent to the curve at this point make with the positive direction of the x axis?

7 Find the two points of intersection of the curve $y=3x^2-1$ and the straight line $y=x+1$. Find the gradients of the curve at these two points.

8 Find the point on the curve $y=3-4x+2x^2$ at which $dy/dx=0$. Write down the equation of the tangent to the curve at this point.

✳ *9* Find the gradient of the curve $y=x^3-4x^2+x$ at the point A where $x=2$. Find the co-ordinates of a second point where the tangent to the curve would be parallel to the tangent at A.

10 Find the co-ordinates of the points on the curve $y=1+3x^2-x^3$ where $dy/dx=0$. Hence write down the equations of two tangents to the curve which are parallel to the x axis.

11 Find the gradient of the curve $y=x^2-2x$ at the point where $x=0$. Hence find the equation of the tangent at this point.

180

12 *a*) Find the co-ordinates of the two points A and B at which the curve
$y = x - \dfrac{x^2}{2}$ cuts the x axis.

b) Find the gradients of the curve at the points A and B.

c) The tangents to the curve at A and B intersect at C. Write down the equations of the tangents and hence find the co-ordinates of C.

Velocity and Acceleration

13 We have already seen (in Book 3) that the velocity of a moving body at time t can be found from a distance/time curve by measuring the slope of the tangent at time t.

a) If s is the distance gone in time t, give a calculus expression for the velocity at time t.

b) We have also seen that the acceleration at time t is given by the slope of the tangent to the velocity/time curve. If the velocity is v, give a calculus expression for the acceleration.

14 A particle is moving in a straight line. Its distance from the starting point is s metres after t seconds, where s is given by the equation $s = 9t - 2t^2$. By differentiating with respect to t, find an expression for the velocity of the particle at time t. Calculate its velocity 1 second after starting. Find also at what time the particle comes to rest.

15 The distance of a particle from its starting point is s metres after t seconds. The value of s is given by the equation $s = 10t + 5t^2$.

a) Differentiate with respect to t and find an expression for the velocity.

b) At what time is the velocity 20 m/s?

c) How far has the particle travelled to reach a speed of 20 m/s?

16 The velocity of a particle moving in a straight line is given by $v = 2 + 3t$, where v is the velocity in m/s at time t seconds after the start of the motion. By differentiating with respect to t, find an expression for the acceleration of the particle at any time. What can you say about the acceleration?

17 The equation $v = t^2 - 5t$ gives the velocity of a particle in m/s after t seconds. Differentiate with respect to t and find an expression for the acceleration after time t.

a) After what time is the acceleration 0?

b) What is the velocity when the acceleration is 0?

c) At what time is the particle at rest?

21C Maximum and Minimum Values

1 The graph shows a sketch of the curve
$y = x^2 - 2x - 3$.

a) If A and B are the points of intersection with $y = 0$, find the co-ordinates of A and B.

b) Differentiate with respect to x and find the gradient of the curve at any point (x, y).

c) Find the co-ordinates of the point C where $dy/dx = 0$.

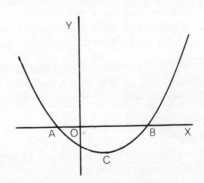

d) What do you know about the tangent at C?

e) Is the value of y a maximum or a minimum at C? (Refer to the sketch.)

f) Find the gradients of the tangents at A and B.

g) Is the gradient of the curve increasing or decreasing as x moves from A to C to B?

2 Draw a sketch of the curve $y=6-x-x^2$ and repeat question 1.

3 *a)* For the curve $y=x^2-4x+3$, find the value of x for which $dy/dx=0$.

b) By taking a value of x before the turning point and another value of x after the turning point, find whether dy/dx is increasing or decreasing with increasing x.

c) Is the turning point a maximum or a minimum? What is the value of y at this point?

4 Find the turning points on the following curves and determine whether they are maxima or minima:

a) $y=x^2-3x$ *b)* $y=x(4-x)$ *c)* $y=x^2+5x+4$

d) $y=2x^2-6x+3$ *e)* $y=x(5-3x)$.

5 *a)* Find the values of x for which $dy/dx=0$ for the curve $y=x^3-3x^2+4$.

b) Find the sign of dy/dx at three points on the curve, one before the first turning point, the second one between the two turning points, and the third after the second turning point. Hence determine which point is a maximum and which a minimum, and calculate the maximum and minimum value of y. Draw a rough sketch of the curve.

6 Find the co-ordinates of the two turning points on the curve $y=x^3-2x^2+x$. By finding the sign of dy/dx at three other points on the curve, determine which turning point is a maximum and which is a minimum.

7 Find the points on the following curves where $dy/dx=0$. Determine whether the points are maxima or minima by finding the sign of dy/dx before and after each turning point. Draw a rough sketch of each curve.

a) $y=x^2(x+3)$ *b)* $y=12x-x^3$ *c)* $y=x+\dfrac{1}{x}$

d) $y=\dfrac{x^3}{3}+\dfrac{x^2}{2}-2x+1$ *e)* $y=2+2x^2-\dfrac{x^3}{3}$

8 *a)* If $s=18t-5t^2$, find the maximum value of s.

b) A particle is projected vertically upwards. After t seconds its distance above its starting point is s metres, where $s=18t-5t^2$. Find the velocity ds/dt and the time at which this is zero. What height has it reached at this time?

c) Are your answers to a and b the same?

Note In a you are carrying out an 'abstract' calculation and finding the maximum value of s by putting ds/dt equal to 0. In b you are giving physical meanings to s and t and saying that s is a maximum when the velocity ds/dt is 0.

9 The perimeter of a rectangular plot of land is 200 m. If the length is x metres, write down an expression for the width. Show that A, the area of the plot, is given by $A=x(100-x)$ where A is in m². Hence find the dimensions of the rectangle for which A is a maximum.

10 The area of a rectangular playground is $3600 \, \text{m}^2$. If the length is x metres, write down an equation for P, the perimeter of the playground, in metres. Find the dimensions of the rectangle for which P has its least value. What is this value?

11 Mr O'Flaherty has $80 \, \text{m}$ of fencing with which to make a rectangular enclosure for his two donkeys. If an existing wall forms one side of the enclosure, write down an expression for the length of the side opposite the wall, using x for the length in metres of each of the other two sides. Find x when the area enclosed is a maximum and find also this maximum area.

12 The length of a cuboid is twice its width and the volume enclosed is $1536 \, \text{cm}^3$. If $x \, \text{cm}$ is the width, write down an expression for the height of the cuboid. Hence find the dimensions of the cuboid for which the sum of the lengths of its edges is least.

13 A small box has a base which is a square of side $x \, \text{cm}$. The sum of its height and the length of one side of the base is $21 \, \text{cm}$. Write down an expression for its volume. Hence calculate x when the volume is a maximum and find this volume.

✳14 The total surface area of a solid cylindrical wooden block is $288 \, \text{cm}^2$. Find an expression for the height of the cylinder in terms of the radius and π. Then find an expression for the volume of the wood in terms of the radius, and hence find the radius of the cylinder which has the largest possible volume. What is the height of this cylinder and what is its volume? (Take the approximate value of 3 for π.)

15 A particle starts from rest at O and moves along a straight line until it comes to rest at P. After t seconds it has velocity v, given by the equation $v = 4t - 3t^3$. Calculate

a) its velocity after $0 \cdot 5$ second
b) the time taken to reach P
c) the time taken to reach maximum velocity
d) the maximum velocity.

21D Integration

1 From the previous exercises we have seen that if $y = x^2 + 3$, differentiation gives $dy/dx = 2x$. Similarly if $y = x^2 + 5$ or $y = x^2 + 8$, $dy/dx = 2x$. The inverse process, integration, written $\int 2x\,dx$, gives $x^2 + A$, where A is a constant that can have any numerical value, e.g. 3, 5 or 8, etc. The value of A cannot be determined without further information. A is called the *constant of integration* or the *arbitrary constant*. Write down the values of:

a) $\int x\,dx$ *b)* $\int 3x\,dx$ *c)* $\int \dfrac{x}{2}\,dx$ *d)* $\frac{1}{2}\int x\,dx$.

2 If $y=x^3+x+2$, then $dy/dx=3x^2+1$. Write down the values of:

a) $\int x^2dx$ b) $\int 2x^2dx$ c) $\int 1dx$ d) $\int 2dx$.

Check that in each case the differential coefficient of your answer is the same as the function you are starting with.

3 If $y=x^n$, then $dy/dx=nx^{n-1}$. Write down the values of:

a) $\int nx^{n-1}dx$ b) $\int x^{n-1}dx$ c) $\int x^ndx$.

d) The integral you have worked out in c does *not* apply when $n=-1$. Can you see why not? Finding the integral of $1/x$ is beyond the scope of this book.

4 Write down the integrals of the following expressions:

a) $3x^2-5$ b) $2x+3$ c) $4x-9x^2+2x^3$ d) $1-4x^3$

e) $x(6-x)$ f) $(4-x)(2-x)$ g) $\dfrac{1}{x^2}$ h) $\dfrac{5x-3x^2+4x^3}{x}$

i) $\dfrac{3-5x^2-3x^4}{x^2}$ j) $\dfrac{1}{x^3}-\dfrac{1}{x^4}$

5 Integrate the following with respect to x:

a) x^5 b) $x^{\frac{1}{2}}$ c) $x^{\frac{1}{3}}$ d) x^{-3} e) $3x^{-2}$

6 Find the values of the following integrals:

a) $\int x^6dx$ b) $\int\left(x^3-\dfrac{x^2}{3}\right)dx$ c) $\int x(1-3x)dx$ d) $\int(6x^{-2}-4x^{-3})dx$ e) $\int\left(x+\dfrac{1}{x}\right)^2dx$.

7 If $dy/dx=9x^2-2x$ and $y=1$ when $x=1$, find y in terms of x.

8 Given that $dy/dx=x^2-x$, and $y=1$ when $x=0$, find y in terms of x.

9 If $ds/dt=3t^2-4$ and $s=3$ when $t=2$, find s in terms of t.

10 If $dv/dt=2t^2-3t+1$ and $v=5$ when $t=3$, find v in terms of t.

11 The gradient of a curve at any point is given by $dy/dx=3x-2$. If the curve goes through the origin, find its equation.

12 A curve which passes through the point $(1, 2)$ has a gradient of $4x+1$ at any point. Find the equation of the curve.

13 The gradient at any point on a curve which passes through the point $(3, -8)$ is given by $dy/dx=5-6x$. Find the equation of the curve.

14 The gradient of a curve at any point is $3x^2+4x-5$. If this curve passes through the point $(-1, 4)$ find its equation.

15 A particle moves from rest along a straight line. Its displacement along this line after t seconds is s metres, and its velocity in metres per second at that time is given by $ds/dt=6t-7$. If after 4 seconds $s=10$ metres, find an equation which relates s to t.

16 The acceleration of a moving particle is given in terms of t by the equation $dv/dt=t(1-3t)$. When $t=0$ the particle is at rest and $s=0$. Find an equation for v in terms of t and hence find an equation for s in terms of t.

21E Definite Integration

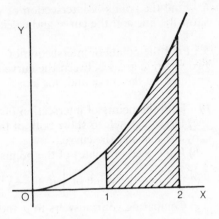

1 The diagram shows the curve $y=x^2$. The shaded area under the curve between $x=1$ and $x=2$ is given by the definite integral

$$\int_1^2 x^2 dx.$$

To evaluate the integral, x^2 is integrated in the usual way *leaving out* the arbitrary constant C. The terms are written in a square bracket with the limits as shown. The value of the terms in the bracket is then calculated first for $x=2$ and then $x=1$, and the answers subtracted. This gives the numerical value of the *definite integral*.

$$\int_1^2 x^2 dx = \left[\frac{x^3}{3}\right]_1^2 = \frac{2^3}{3} - \frac{1^3}{3} =$$

a) Copy and complete the calculation. What is the shaded area?
b) Work out the terms in the square bracket again, this time including the arbitrary constant C. What do you get? Why is C normally left out?

Note If you like 'digging deeper' and are wondering why the integral stated represents the area shaded and why the integral is evaluated in the way stated above, look at question 11.

2 Evaluate the following definite integrals:

a) $\int_0^3 (3x^2 - 2x)dx$ *b)* $\int_2^4 (6x^2 - 1)dx$ *c)* $\int_1^4 x(4-9x)dx$

d) $\int_1^4 \left(4 - \frac{1}{x^2}\right)dx$ *e)* $\int_1^5 (3-2x)dx$ *f)* $\int_1^2 \frac{2}{x^3}dx$

g) $\int_{-2}^{-1} \left(x + \frac{1}{x^2}\right)dx$ *h)* $\int_0^2 (x+2)(x+3)dx$

i) $\int_{-1}^1 x^2(4x-3)dx$ *j)* $\int_{-3}^{-1} \frac{3x^2 - 1}{x^2}dx$

3 Draw a sketch of the curve $y=x^2 - 2x - 3$. Find the area between $y=0$ and the curve from $x=-1$ to $x=3$.

4 Find the two points at which $y=x^2 - 4$ cuts the x axis. Hence calculate the area enclosed.

5 Draw a sketch of the curve $y=6-x-x^2$. Calculate the area enclosed by the curve and the x axis.

6 Find the area bounded by the two axes, the curve $y=x^2 + 4x + 5$ and the line $x=2$.

7 Draw a sketch showing the curve $y=x^2 - 4x + 6$ and the line $y=6$. By subtracting areas, calculate the area enclosed by the line and the curve.

185

8 Find the points of intersection of $y=x-2$ and $y=x^2-x-2$. Draw a sketch to show the line and the curve and calculate the area between them.

9 Find the points of intersection of $y=x^3-5x^2+6x$ with the x axis. Calculate the two separate areas between the curve and the x axis. What is the total area which is enclosed by the curve and this axis?

10 Find the points of intersection of $y=8+2x-x^2$ and $y=x+2$.
  a) Draw a sketch to show both of these and hence calculate the finite area bounded by the line and the curve.
  b) If the x co-ordinates of the points of intersection are denoted by c and d,

  $(d>c)$, find the value of $\displaystyle\int_c^d (6+x-x^2)dx$.

  c) Comparing your answers to a and b what do you notice?

* 11 *The Fundamental Theorem of the Calculus*
Consider the area under a curve, bounded by the curve, the ordinate $x=a$, the x axis and an ordinate through the point (x, y). Let this area be A. Then A is a variable. Draw a narrow strip of area of width δx. Call the area of this strip δA. Then to a first approximation $\delta A = y\delta x$. The strip is shown in detail in the second sketch. The approximate area between the ordinates $x=a$ and $x=b$ could be thought of as the sum of a large number of these little strips, and could be written as

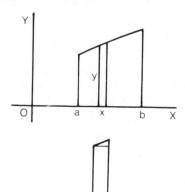

$$\sum_{x=a}^{x=b} y\delta x$$

where Σ is a Greek capital S (capital sigma) denoting summation.
The larger the number of strips, and the smaller δx, the closer we get to the true area. So we can write

$$A = \underset{\delta x \to 0}{\text{Lt}} \sum_{x=a}^{x=b} y\delta x$$

which in words means that A is the limit as δx approaches zero of the sum of the products $y\delta x$ between $x=a$ and $x=b$. This clumsy notation is now replaced by the more elegant notation $\displaystyle\int_a^b ydx$.

So now we have a notation for the area, but we need to find a meaning for the notation. Going back to the statement $\delta A \simeq y\delta x$ (where $\simeq$ means 'is approximately equal to') and dividing both sides by δx, we get $\delta A/\delta x \simeq y$. Proceeding to the limit where $\delta x \to 0$, we get $dA/dx = y$. This gives us what we need.

We can now say that $A = \displaystyle\int_a^b ydx$ where A is such a function that its differential coefficient is y. In other words, integration is the reverse of differentiation as stated in question 1.

  a) Can you see one enormous assumption in the above 'proof'?

b) Can you add a piece to the 'proof' to show that the numerical value of the integral is $\left[Y \right]_a^b$ where Y is the integral of y?

c) Can you draw one or more curves between $x=a$ and $x=b$ for which the 'proof' might break down?

Note The above theorem is called *The Fundamental Theorem of the Calculus* since it relates the differential to the integral calculus.

21F Miscellaneous

1 Find the gradient of the curve $y=2x^2-5x+3$ at any point (x, y). Calculate the co-ordinates of the point on the curve at which the tangent to the curve is parallel to the line $y=3x$.

2 The tangent to the curve $y=ax^2+bx+6$ at the point $(3, 0)$ is parallel to the line $y=x$. Find the values of a and b and the co-ordinates of the point on the curve at which the tangent is parallel to $y=0$.

3 Find the co-ordinates of the points on the graph of the function $y=x^3-x^2-5x+2$ where y is a maximum or a minimum. For what range of values of x is dy/dx negative?

4 Draw a sketch of the curve $y=x(x-3)$. Find the finite area bounded by the x axis and the curve. Show that the area between the curve and the x axis and $x=-3$ is five times this area.

5 Calculate the co-ordinates of the points on the curve $y=3x^3-2x+1$ where the tangents to the curve are parallel to $y=2x$.

6 Calculate the maximum and minimum values of y if $y=x^3+2x^2-4x+5$. Find also the co-ordinates of the point on the curve at which the gradient is a minimum.

7 The velocity v of a particle moving in a straight line is given by $v=(t-2)(t-5)$, where t is the time. Find an equation for s, the distance travelled, in terms of t, given that $s=0$ when $t=0$. Hence find the distance travelled between the two instants when the particle is at rest.
Find also the accelerations of the particle when it is at rest.

8 The curve $y=x(4-x)$ cuts the x axis at A and B. The x co-ordinate of A is less than that of B. The tangent to the curve at B cuts the y axis at C. Calculate the area bounded by the arc AB and the lines AC and BC.

9 A and B are the points on $y=3x^2+\dfrac{2}{x}$ whose x co-ordinates are -1 and 1 respectively. Find the gradient of the curve at A and the gradient of the line AB.

10 The velocity of a particle moving in a straight line is given by $v=(t-4)(t-2)$, where t is the time.

a) Find the times at which the velocity is 3.
b) Draw a sketch of the curve and find the distance travelled between these two times.
c) Why is the distance so small?

11 Sketch the curve $y = \dfrac{1}{x^2}$ for the range $1 \leqslant x \leqslant 4$. Find the area bounded by the curve, $y = 0$, $x = 1$ and $x = 3$.

12 An outside contractor is engaged to carry out urgent repairs to a tunnel. If his bricklayers work overtime the work is speeded up and his overheads are reduced, but he has to pay a higher rate for the overtime worked. The cost of the materials of course is unaffected. An engineer in his office estimates that C, the cost of the entire job in thousands of pounds, is given by the equation $C = 7(h-8)(h-12) + 600$ where h is the number of hours worked in a day.

   a) Find the value of h that makes C a minimum.
   b) What is the minimum total cost?
   c) There is a penalty clause in the contract agreement by which the contractor loses £1000 for every day by which he exceeds the agreed time for the job. He decides to work 11 hours a day and make sure of completing it in time. By how much will the total costs now exceed the minimum?

13 A racing track is in the shape of a rectangle of length l metres with a semi-circle of radius r metres at each end. The outside perimeter of the track is 2000 metres.

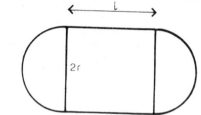

   a) Write down an equation in l and r and hence express l in terms of r.
   b) Write down the area A in terms of l and r, and using the results of *a* express A in terms of r only.
   c) Find the maximum value of A and the corresponding values of l and r.

(*Hint* Retain π in your calculations till the final answer.)

14 Repeat 12 for a track with a semi-circle on one end only.

15 For the first six hours after a heavy and prolonged thunderstorm, water is pouring into a reservoir at a rate given approximately by the equation $Q = 10 + 24t + 3t^2 - t^3$, where t is the time in hours after the start of the storm and Q is the inflow in thousands of cubic metres per hour.

   a) Find the maximum rate of inflow and the time at which it occurs.
   b) The area of the surface of the reservoir is 80 hectares. At the peak period water is flowing out of the reservoir at the rate of 10 000 cubic metres per hour. At what rate is the water level rising?
   c) What is the total volume of water that flows into the reservoir during the first six hours after the start of the storm?
   d) If during this period the average rate of outflow is 5000 cubic metres per hour, by how much does the water level rise altogether?

✳*16* A batch of n articles costs C pence per article to produce. Part of the cost C varies directly as n and part varies inversely as n.

   a) Using the constants p and q, write an equation connecting C and n.
   b) If C is a minimum when $n = 100$, find a relation between p and q.
   c) If the cost per article is £3·03 when $n = 10$, using the results of *b* find the values of p and q.
   d) What is the cost per article in a batch of 50?
   e) What is the minimum cost per article?

*17 Repeat question 16 if part of the cost C varies directly as n and part varies inversely as n^2, the minimum value of C occurs when $n=10$ and the cost per article when $n=5$ is £2·50.

18 *a)* Draw a graph of the parabola $y=x^2+4x+6$ between $x=0$ and $x=6$.
     b) Find by integration the area under the curve.
     c) Show that for this parabola Simpson's rule gives the exact area.

19 Repeat question 18 for the cubic $y=x^3-9x^2+24x$. Does Simpson's rule give an exact answer?

20 Repeat question 18 for the quartic $y=x^4-5x^3-8x^2+64x$. Does Simpson's rule give an exact answer this time?

21 The sides of a right-angled triangle are x and y.

    a) Write down the area A in terms of x and y.
    b) If $x+y=10$, write A in terms of x only.
    c) Find the minimum value of A and the corresponding values of x and y.
    d) How could you have deduced the values of x and y without using the calculus?

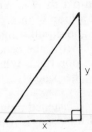

22 *a)* A rectangle has a fixed perimeter. What shape must it be to enclose the maximum area? (Refer back to 21C questions 9 and 10, or work it out afresh using the calculus.)
    b) The sum of the lengths of all the edges of a cuboid is 120 cm. If one of the faces is rectangular with two unequal edges, how could you increase the volume of the cuboid, still keeping the sum of the lengths of the edges at 120 cm?
    c) If after you have done this another face is rectangular with unequal sides, could you increase the volume again without altering the sum of the lengths of the edges?
    d) Repeating this process indefinitely, what would you end up with?
    e) What is the maximum possible volume of this cuboid?

22 Miscellaneous Topics

22A   Loci

1   A boy walks along a straight, level path. What is the locus of the tip of his nose?

2   A girl sits on one end of a seesaw, which is then set in motion. Sketch the locus of the tip of her nose.

3   A goat is tethered by a light chain to an iron stake on the grass verge of a road. If he walks round the post keeping his chain tight all the way, what is his locus?

4   Stick two pins into a sheet of paper on a drawing board. Put a loop of thin string loosely over the pins, and pull it tight with the point of a pencil. Move the pencil round the pins keeping the loop tight. What is the locus of the point of the pencil?

5   A piece of paper is fastened tightly round a circular drum and the drum is rotated slowly about a vertical axis at a steady speed. If the point of a pencil is held stationary against the side of the drum for one complete revolution and the paper then removed, what locus will the pencil point have traced out on the paper?

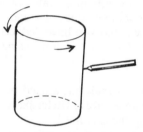

6   Repeat question 5 if the pencil is moved slowly downwards at a steady speed.

7   A ball is thrown vertically upwards. What is its locus?

8   A ball is thrown at an angle to the vertical. Sketch its approximate locus. Do you know the name of the curve?

9   A point moves so that its distance from a fixed point is equal to its distance from a fixed line. By drawing a few positions of the point, sketch its locus. Do you know the name of the curve?

10   In each part of this question, describe the locus of the point P in words and also draw a sketch to illustrate your answer. P moves so that:

a) its distance from a fixed point O is constant
b) its distances from two fixed points are equal
c) its distances from two fixed intersecting lines are equal (two answers)
d) its distances from two fixed parallel lines are equal
e) its distance from one of a pair of parallel lines is double its distance from the other (four answers)
f) the angle APB is 90°, where A and B are fixed points
g) the angle APB is 60°, where A and B are fixed points
✱ *h)* $AP:PB=3:2$ where A and B are fixed points
✱ *i)* O is a fixed point, OA a fixed direction, angle $AOP=a°$, $OP=a/180$, and a is a variable.

*11 A wheel rolls along a straight, level road without slipping. What is the locus of a fixed point on the rim?

This locus is known as a *cycloid*. To draw it either:

    a) Cut a circle in fairly stiff card, make a small notch on the circumference, and roll it flat, without slipping, along a straight line on a piece of paper. Get someone to help you. At intervals make a pencil mark on the paper against the notch. Join up the marks with a smooth curve after the circle has rolled a full revolution.

    b) Draw a series of circles as shown, the distance between their centres being $\frac{1}{8}$ of the circumference of the circle. Mark points on the circumference at angles of 45°, 90°, etc., as shown. Join these points up to give a smooth curve.

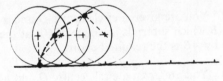

*12 If *a* is the height of one arch of the cycloid in question 9, calculate the area of the arch, using either Simpson's Rule or some other method. Divide this area by $3a^2$. What do you get? What do you think is the correct expression for the area of one arch of a cycloid?

*13 Draw two lines *OX* and *OY* at right angles. Draw a line *AB* 6 cm long, with *A* on *OY* and *B* on *OX*. Draw a large number of positions of *AB*. The envelope of all these lines (i.e. the curve that touches them all) is called a *tractrix*. The longitudinal section of a trumpet is a tractrix. What is the meaning of the name?

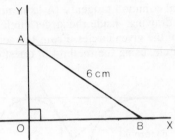

22B Ruler and Compass Constructions

1 To find the mid-point of the line *AB*, with centres *A* and *B* and equal radii, draw two arcs intersecting at *P* above the line *AB*. With centres *A* and *B* and equal radii, draw two arcs intersecting at *Q* below the line *AB*. Join *PQ*. If the line *PQ* cuts *AB* at *X*, then *X* is the mid-point of *AB*. *PQ* is also the perpendicular bisector or mediator of *AB*. In the above construction, state which radii *must* be equal and which *need not* be equal.

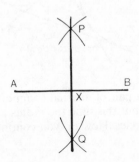

2 Find a ruler and compass construction for drawing a line perpendicular to a given line *YZ* at a given point *X* on *YZ*.
(*Hint* The finished construction will look like the figure in question 1.)

3 Repeat question 2 for a perpendicular to a given line *YZ* from a point *P* outside the given line.

4 Find a ruler and compass construction for drawing the bisector of a given angle *AOB*. Start off by drawing two equal arcs *OP* and *OQ* as shown. There are two bisectors. Draw them both. What is the angle between them?

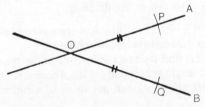

5 Using ruler and compass only, construct the following angles. In each case, give your construction in full and show all construction lines.

a) 60°   *b)* 45°   *c)* 30°   *d)* 22½°   *e)* 15°   *f)* 75°   *g)* 37½°   *h)* 67½°

6 Draw a circle through three points *A, B* and *C* which are not in a straight line. Start by drawing the mediator of *AB*. What must you do next? Explain why your construction works.

7 You are given a circle centre *O* and a point *A* on the circumference. Find a chord *AB* which subtends an angle of 60° at the circumference (two answers). Explain briefly why *AB* is the required chord.

8 You are given a circle centre *O* and a point *P* outside it. Draw a tangent from *P* to touch the circle at *T* (two answers). Your construction must give the exact position of *T*. It is not enough to draw the tangent by eye. Explain briefly why your construction gives the exact position of *T*.

✳9 For two non-intersecting circles there are two external common tangents and two internal common tangents. (A 'common tangent' is a line touching both circles.) Start off by drawing, inside the larger circle, a circle whose radius is the difference of the radii of the given circles. From the centre of the smaller circle draw a tengent to this new circle, using the method of question 8. What next?

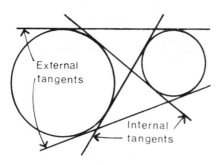
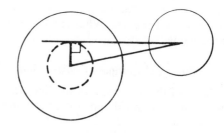

10 Draw a pair of internal common tangents to the circles in question 9. Start as in question 9, but this time the radius of the new circle should be the sum of the radii of the given circles. How do you continue?

22C The Radian Measure of an Angle

1 Angles can be measured in radians as well as in degrees. Radians are used extensively in higher mathematics. A radian is the angle subtended at the centre of a circle of unit radius by an arc of unit length. This is shown in the diagram.

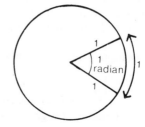

a) A radian is a little less than 60°. How can you tell this from the figure?

b) To find its exact value, we know that the circumference of a unit circle is 2π, so the angle subtended by the above arc is $(360/2\pi)°$, or $(180/\pi)°$, i.e. one radian is $(180/\pi)°$. Work out the value of a radian in degrees correct to 1 d.p.

2 Using the fact that 1 radian=$(180/\pi)$ degrees, calculate the value of the given angles in degrees. Where π does not cancel out, take it as 3·142 and give your answer correct to one d.p.

a) $\dfrac{\pi}{2}$ radians   b) $\dfrac{2\pi}{3}$ radians   c) $\dfrac{\pi}{10}$ radians   d) 1·14 radians

e) $\dfrac{\pi}{4}$ radians   f) 0·85 radians

3 Calculate the values of the following angles in radians correct to 2 d.p.

a) 70°   b) 30°   c) 46°   d) 150°   e) 225°   f) 113°

4 In a circle of radius r, if an arc subtends an angle of θ radians at the centre, then the length of the arc is $r\theta$. In each of the following examples you are given the radius of the circle and the angle at the centre. Find the length of the arc.

a) 6 cm, 2 radians   b) 4 cm, 1·3 radians   c) 5 cm, 38°
d) 2 cm, 1·36 radians   e) 3·4 cm, 57·3°   f) 7 cm, π radians

5 In each of the following examples you are given the radius of a circle and the length of an arc. Using tables of radians, or otherwise, find the angle subtended at the centre *i*) in radians   *ii*) in degrees.

a) 6, 6   b) 3, 5   c) 5, 3   d) 8, 2   e) 7, 4

22D   More about Polar Co-ordinates

Reflections using Polar Co-ordinates

1 P is the point (r, θ), P_1 is the image of P after reflection in the central direction, $\theta=0$. P_2 is the image of P after reflection in $\theta=45°$. Find the polar co-ordinates of P_1 and P_2.

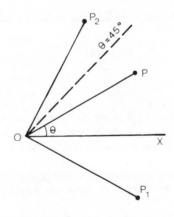

2 Find the images of the given points after reflection in the lines stated.

a) (3, 40°), $\theta=60°$ b) (2, 70°), $\theta=90°$ c) (5, 40°), $\theta=270°$
d) (4·2, 18°), $\theta=70°$.

3 Find the images of the following points after reflection in the lines stated, all angles being expressed in radians.

a) (2, $\pi/4$), $\theta=\pi/2$ b) (3, 2), $\theta=0$ c) (3, $\pi/2$), $\theta=\pi/3$ d) (1·5, $2\pi/3$), $\theta=\pi$

Polar Co-ordinates and Loci

4 What is the polar equation of a circle of radius a?

5 What is the polar equation of the line whose cartesian equation is $x=a$?
(*Hint* If P is a point on the line, what is its x co-ordinate ON in terms of r and θ?)

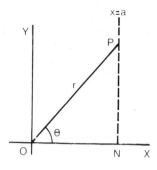

6 Give the cartesian equation of the line whose polar equation is $r\sin\theta=b$.

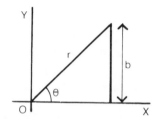

7 Draw three complete whorls of the Archimedes Spiral $r=\theta/2$, where θ is measured in radians.

8 Repeat question 7 for the spiral $r=\theta/120$, where θ is in degrees.

The Limaçon and the Cardioid

9 Copy and complete the following table:

θ (degrees)	$\cos\theta$	$4\cos\theta$	$r=6+4\cos\theta$
0	1	4	10
30	0·87	3·48	9·48
60	0·5	2·0	8·0

Take values of θ every 30° from 0° to 360°, calculating extra values where needed. Taking an origin fairly near the left side of the paper and midway between top and bottom, draw the curve. It is known as a *limaçon*. Its polar equation is $r=6+4\cos\theta$.

10 Repeat question 9 for the limaçon $r=2+4\cos\theta$. What difference do you notice between the curves in questions 9 and 10?

11 Repeat question 9 for the equation $r=4+4\cos\theta$. This time the curve has a cusp at the origin. It is known as a *cardioid* (i.e. heart-shaped curve) but of course it is also a limaçon.

Note The general curve for the limaçon is $r=k+2a\cos\theta$.
When $k>2a$, there is a dimple but no loop. When $k=2a$ there is a cusp and the curve is a cardioid. When $k<2a$ there is a loop.

22E Combining Sine and Cosine Curves

1 On the same axes draw the curves of $y=\sin x$ and $y=\cos x$ from $x=0°$ to $x=360°$. Add the co-ordinates of the two curves and draw a fresh smooth curve through the points so obtained. This new curve should also be a sine curve, but with a different scale and a different origin from the original sine curve. Its equation is $y = \sin x + \cos x$.

2 Copy and complete the following table, working to 2 decimal places and taking values of x from $0°$ to $360°$ in steps of $45°$:

x	$x+45°$	$\sin(x+45°)$	$1\cdot41\sin(x+45°)$
$0°$	$45°$	$0\cdot707$	$1\cdot00$
$45°$	$90°$		

Plot the graph of $y = 1\cdot41\sin(x+45°)$. Compare it with the graph of $y = \sin x + \cos x$ in question 1. What do you notice?

3 Arrange a table with the columns x, $\sin x$, $3\sin x$, $\cos x$, $4\cos x$ and $y = 3\sin x + 4\cos x$. Calculate values of y for a domain of $0°$, $30°$, ... $360°$. Plot the graph of $y = 3\sin x + 4\cos x$ from $x = 0°$ to $x = 360°$.

4 Draw up a table for calculating the values of $5\sin(x+53)°$ from $x=0$ to $x=360$ by steps of $30°$. Draw the curve $y = 5\sin(x+53)°$ and compare it with the curve in question 3. What do you notice?

*5 Questions 1 to 4 are examples of a remarkable fact, that when sine and cosine curves are combined (either the simple curves or multiples of them) they give another sine curve. This can be expressed mathematically as

$a\sin x + b\cos x = c\sin(x+d)$ where $c = \sqrt{a^2+b^2}$ and $\tan d = b/a$.

a) Taking $a=5$, $b=12$, i) calculate the values of c and d, ii) check that this formula is true for $x=20°$, iii) check also that it is true for $x=50°$.
b) Repeat a) for $a=2$, $b=3$.

*6 Using the formula in question 5, $2\sin x + 5\cos x$ should be equal to $\sqrt{29}\sin(x+\tan^{-1}(5/2))$, i.e. $5\cdot39\sin(x+68\cdot2°)$.
Draw up a set of tables for $y = 2\sin x + 5\cos x$ and $y = 5\cdot39\sin(x+68\cdot2°)$, calculating values at intervals of $30°$ from $0°$ to $360°$. You should get the same values for both curves.

22F Open and Closed Intervals on the Number Line

Finite Intervals

1 Consider the intervals:
    i) $3 \leqslant x \leqslant 8$ ii) $3 \leqslant x < 8$ iii) $3 < x \leqslant 8$ iv) $3 < x < 8$
A finite interval that contains both end points is known as a closed interval. A finite interval that does not contain both end points is known as open. (Some writers use the name 'half open' for a finite interval which includes one end point only.) State which of the above intervals are open and which are closed.

2 Intervals can also be represented by using square brackets. Thus $]2, 6]$ represents the open interval $2 < x \leqslant 6$. The brackets $]3, 7]$ indicate that 3 is excluded from the interval and 7 is included. The brackets $[3, 7[$ indicate that 3 is included and 7 is not. Represent the four intervals in question 1 in this way.

3 If the domain is the reals, the finite intervals *i*) to *iv*) in question 1 can be represented on the number line as shown, where ● indicates that the number is included in the interval and ○ indicates that the number is not included in the interval.

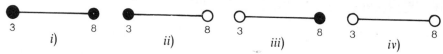

If the domain is the integers, the representation would be:

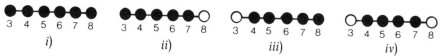

Represent the following sets on the number line, the domain being indicated as *I* (integers) or *R* (reals).

a) $-2 \leqslant x < 4$ (*I*)   b) $-2 < x \leqslant 6$ (*R*)   c) $4 < x \leqslant 9$ (*I*)
d) $0 \leqslant x \leqslant 5$ (*R*)   e) $2 \leqslant x < 7$ (*R*)   f) $-5 < x < 1$ (*I*)
g) $7 > x > -1$ (*R*)   h) $3 \geqslant x \geqslant -4$ (*I*)

4 Represent the two intervals a) $4 \leqslant x \leqslant 7$ and b) $3 < x < 8$ in the integral domain on the number line. Are the two intervals the same? Are the sets of numbers in the intervals the same?

Non-finite Intervals

5 $x \geqslant 3$ is a non-finite interval and so is $x < 6$ or $x \leqslant -4$. A non-finite interval is closed if it contains the one end point. It is open if it does not contain the end point. (Some writers however call all non-finite intervals open as one end of the interval is always unbounded.) State which of the following non-finite intervals are closed and which are open:

a) $x \leqslant 7$ b) $4 > x$ c) $-2 \leqslant x$ d) $x > -1$ e) $2 \geqslant x$

6 Represent the following non-finite intervals on the number line:

a) $5 \geqslant x$ (*R*)   b) $-3 < x$ (*I*)   c) $x < 6$ (*R*)
d) $x \geqslant -2$ (*I*)   e) $x \leqslant -5$ (*R*)   f) $4 \leqslant x$ (*R*)

23 Everyday Arithmetic and Enlarged Horizons

23A Everyday Arithmetic

Note The details given in this section are realistic at the time of going to press, but will need up-dating regularly.

Take-home Pay

1 A factory worker earns £100 in a certain week. From this the following deductions are made:

National Insurance	£6·52
Pension	£6·00
Income Tax	£10·40
Voluntary Savings	£2·00
Holiday Fund	£4·00

a) Work out his 'take-home pay'.

✱*b)* If the employer pays simple interest at the rate of 0·15p per week on every pound paid into the holiday fund, find how much the above worker would draw at the end of 49 weeks if he had paid in £4 every week for the full 49 weeks. (*Hint* Multiply the total interest earned by the £4 deposited in the middle week by the number of weeks. This gives the total interest added.)

2 *a)* The employer also pays a substantial contribution into the National Insurance Fund. Thus on a gross pay of £60 in one week, the employee's contribution is £3·92 and the employer's contribution is £8·13. Assuming there are 25 million employed persons in Great Britain and that the average wage is £60 a week, what would be the total sum paid into the National Insurance in a full year? Give your answer to the nearest billion pounds.
b) How is this vast sum spent?

3 An employee is paid £1·50 an hour for a 38-hour week. The first five hours of overtime are paid at 'time and one third' and any further overtime worked is paid at 'time and two-thirds'.
a) If he works fifty hours in a certain week, what is his gross pay that week? If the income tax deducted is £7·10 and the National Insurance contribution is £5·49 and there are no voluntary savings or holiday payments, what is his take-home pay?
b) If he works 54 hours the next week, his income tax deduction goes up by 30p on every extra pound earned, and his National Insurance payment is £6·16, what is his take-home pay?

4 A lady member of the staff of a firm is paid monthly and her annual salary is £5100. Superannuation is deducted at the rate of 6% of her gross annual salary. Her total allowances for income tax purposes come to £2000 a year, and the remainder of her salary is taxed at 31p in the £. National Insurance deductions are £27·69 a month. If she contributes £15 a month to the holiday fund and this earns interest at the rate 0·65p per £ per month, and she also pays £10 a month in voluntary savings, answer the following questions.
a) What is her gross monthly pay?
b) What is the monthly income tax deduction?
c) What is her net monthly pay?
d) What sum does she receive for holiday savings at the end of eleven months?
(*Hint* Multiply the interest on the £15 paid in the middle month by 11.)

5 A man's take-home pay is £86 a week. He gives his wife £30 a week for housekeeping and personal expenses, and £26 a quarter for the children's clothes and shoes. Coal for the central heating costs £300 a year, electricity £28 a quarter, rates £110 a year, insurances (life, house and contents) £120 a year, mortgage repayments £10 a week, car expenses, including road fund licence, hire purchase payments, repairs and maintenance but excluding petrol, £500 a year, petrol £4 a week. He also puts £5 a week into the Post Office Savings Bank to pay the cost of the summer holiday. Taking a year as 50 weeks, a quarter as 13 weeks and working each separate cost to the nearest £1 a week, if he puts aside £5 a week to cover emergency expenses, how much does this leave him for personal expenses?

6 Another lower paid worker has a take-home pay of £50 a week. He gives half of this to his wife for housekeeping and the children's clothes and her own personal expenses. She finds it necessary to supplement this by doing a part-time job. Of the remaining money he spends £200 a year on coal, £40 a quarter (average) on electricity, £9 a week on rent and rates for his council house and £2 a week on life insurance. He does not run a car and goes to work on a bicycle. The money that is left goes to pay emergency expenses and his necessary personal spending. He has to pay an unexpected bill of £16. He spreads payment over four weeks. How much is left for his personal expenses per week during that four-week period?

7 A young couple are saving to set up a home. Between them they have a take-home pay of £130 a week. They live in a furnished flat for which they pay £25 a week. Living carefully they keep their personal expenses to £5 a week each, plus £35 a week for heat, light, food, laundry, etc. They own a small car which is fully paid, and they keep the cost of running this (including petrol) down to £10 a week all in. Life insurance for both of them costs £3 a week, and they allow holiday expenses of £200 a year. They deposit the rest of their money in a building society and decide to buy a house when their savings amount to £3000, intending to pay £2000 down on the house and £1000 on furnishings, the balance being paid by mortgage and hire purchase. How long will it take them to save the £3000 they need, ignoring the interest they are paid on their deposits by the building society?

8 The young couple in question 7 find a house on which they will have to pay £2000 down and the mortgage repayments will be £100 a month for 30 years. The rates will amount to £100 a year. The house is in good repair. They choose their furniture and decide to pay £1000 down and £75 a month hire purchase for 3 years. Can they afford to go ahead and buy the house and furnishings?

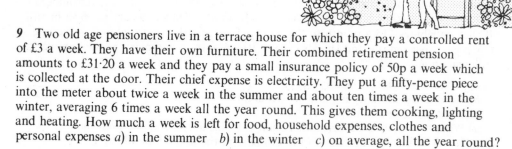

9 Two old age pensioners live in a terrace house for which they pay a controlled rent of £3 a week. They have their own furniture. Their combined retirement pension amounts to £31·20 a week and they pay a small insurance policy of 50p a week which is collected at the door. Their chief expense is electricity. They put a fifty-pence piece into the meter about twice a week in the summer and about ten times a week in the winter, averaging 6 times a week all the year round. This gives them cooking, lighting and heating. How much a week is left for food, household expenses, clothes and personal expenses *a*) in the summer   *b*) in the winter   *c*) on average, all the year round?

Life Insurance

10 A young man of 25 decides to insure his life for £2500. He takes out a 'whole life' policy for which the annual premium is £20·60 per £1000 cover.

   a) What is the annual premium he must pay?

   b) How much is this per month?

   c) Taking the year as 52 weeks, how much a week does his life insurance cost him?

11 Repeat question 10 for a man of 40, taking out an endowment policy for £2000 at an annual premium of £46·10 per £1000.

12 Repeat question 10 for a man of 55 taking out an endowment policy for £1500 at an annual premium of £114·50 per £1000.

13 As the expectation of life for a woman is slightly greater than for a man, insurance premiums for women are slightly lower than for men. Thus for a woman of 40 the cost per £1000 of an endowment policy would be £45·10 per annum (compared with the £46·10 quoted for a man above). Repeat question 10 for a woman of 40 taking out an endowment policy for £2200.

14 In times of inflation the true value of the sum insured steadily diminishes from year to year. To counter this, most insurance companies give a 'bonus'. Thus for a whole life policy taken out ten years ago for £1000 nominal value, there might be a bonus of £550, so that the total sum received on death would be £1550.
Answer the following, using the figures quoted in question 10.

   a) If the young man dies after 10 years, what is the total sum of the contributions he has paid? If the bonus is £550 per £1000, what is the actual death benefit paid?

   b) If he dies after 20 years, what is the total sum of the contributions he has paid, and what is the death benefit, if the bonus is £1100 per £1000 insured?

15 With a 'whole life' insurance policy, the benefit is paid on the death of the insured person. With an 'endowment' policy, the benefit is paid either at the end of a fixed term of years or on the death of the insured if this occurs before the end of the fixed term of years.

   a) If the endowment policy in question 11 was for 25 years, what would have been the total premiums paid over the whole 25 years?

   b) If the bonus was at the rate of £1400 per £1000 insured, what would be the actual payment made on maturity?

16 In question 12, if the endowment policy was for 10 years and the bonus was £510 per £1000 insured, what would have been *a)* the total premiums paid over 10 years *b)* the actual payment made on maturity?

*17 In questions 15 and 16 the sum repaid is considerably greater than the premiums paid. However, the premiums should also be adjusted for inflation. Thus for an insurance policy for £1000 taken out ten years ago at a monthly premium of £4, if the cost of living has doubled over ten years, by present-day standards the original monthly premium would now be equivalent to £8. Taking the average value of the monthly premium over the ten years as the mean of £4 and £8, i.e. £6, the present-day value of the premiums paid would be 120 × £6, i.e. £720, but the sum paid on death would still be £1550.

   a) For a policy taken out 20 years ago for £1000 at a monthly premium of £3, if the cost of living is 3 times greater today than 20 years ago, what is the present-day value of the total premiums paid? If the bonus is £1100 per £1000 insured, what is the sum payable on death?

b) Repeat question *a* for a policy for £1000 taken out 30 years ago, at a monthly premium of £1·80, if the cost of living is now five times greater, and the bonus is £1650 per £1000 insured.

18 After premiums have been paid for a few years, an insurance policy can be 'surrendered' at any time and a 'surrender' value is paid. For the policy quoted in question 11, the surrender value after 10 years would be £563 per £1000 insured.

a) What is the total value of the premiums paid over the ten years?
b) If the cost of living has doubled, what is the 'adjusted' value of these premiums?
c) What is the surrender value of the policy?
d) Which is the greater, the 'adjusted' value or the surrender value?

19 A man has an insurance policy for a nominal £2000 which he took out 10 years ago at a monthly premium of £4 per £1000 insured. He is in urgent need of capital. The surrender value of the policy is £560 per £1000 originally insured.

a) If he surrenders the policy, what capital sum does he receive?
b) If he deposits the policy with the bank, they are prepared to lend him up to 80% of the surrender value. How much is this?
c) If the bank charges him 12% per annum on the loan, how much interest would he pay per month?
d) If he surrenders the policy and takes out another policy for the same amount, the premium will now be £7 per month per £1000 assured. How much extra premium would he have to pay each month?
e) If he dies ten years after taking out the new policy and there is a bonus of £700 per £1000 insured, how much would his widow get? How much would she have received under the old policy, which would then have been twenty years old, with a bonus of £1300 per £1000 insured?

Insurance-linked Mortgages or Endowment Mortgages

20 A man of 25 buys a house for £15000, pays £3000 down and takes the balance on a 20-year insurance-linked mortgage.

a) If he pays the building society interest at 9·75% on the full amount of the loan, what is his gross monthly payment for interest?
b) If tax relief is allowed at 30% on this interest, what is the net monthly payment for interest?
c) If the charge for the endowment policy is £53·34 per annum on every £1000 assured, what is the gross monthly payment for insurance?
d) If tax relief is allowed on this at 15%, what is the net monthly payment for insurance?
e) What is the total monthly payment? (This is paid in a single sum, not in parts.)
f) No repayment of principal is made throughout the period of the mortgage but on maturity the building society is paid in full by the insurance company, and the man receives any bonus due. If this amounts to £1350 per £1000 originally insured, how much bonus will he receive altogether?
g) What do his total monthly repayments over the 20 years amount to altogether?
h) How much has his mortgage actually cost him?
i) If he dies after ten years and his widow survives him, the mortgage is paid in full

and she receives a bonus. If this amounts to £550 per £1000 originally insured, how much does she get?

Note Bonus payments are not guaranteed. The figures quoted are in line with current payments.

21 How can insurance companies regularly pay back more than they receive in premiums?

Income Tax Allowances

22 Income tax at a standard rate is levied on all earned incomes. However, this standard rate is not charged on the full income as certain 'allowances' are granted and these are deducted from the 'gross income' to give the 'taxable income'. Here is an example:
Gross income, £5200 p.a.; personal allowance (married man), £1535; allowance for life-insurance premiums (in some cases the full annual premiums), £50; allowance for interest on bank loan for home improvements, £70; allowance for mortgage interest paid to a building society, £500.

a) What is the total allowance?
b) What is the taxable income, i.e. the income remaining after all the allowances have been deducted?
c) If tax is levied on the taxable income at 30% p.a., what is the tax for *i*) one year, *ii*) one month, *iii*) one week (take 52 weeks in the year)?
Give all answers to the nearest ten pence.

23 A single man earning £4500 p.a. has the following allowances: personal, £985; life insurance, £78; expenses incurred in pursuit of his business, £56. If the tax rate is 31p in the £, calculate the tax he pays a) per annum, b) per month, c) per week, giving all your answers to the nearest 10p.

24 Repeat question 23 for a widower with an income of £5800, personal allowance £985, additional personal allowance £550, allowance for mortgage interest £620, expenses incurred in the pursuit of his business £84.

25 Repeat question 23 for a single woman who is earning £6000 p.a. She is buying her own house and the allowance for mortgage interest is £730, her personal allowance is £985, allowance for life insurance is £144, an allowance for necessary travelling expenses incurred in carrying out her work is £117, and dependent relative's allowance is £145.

26 A wife's earnings can be taxed separately from her husband's, or the tax can be deducted for both of them from the husband's income. A young couple opt for the latter choice.

a) The husband is earning £75 a week and the wife £25 (part time). What is their joint annual income?
b) They receive the following allowances: personal allowance (married man), £1535; wife's earned income allowance, £985; life insurance, £75; dependent relative's allowance, £100. They are buying a house on an option mortgage scheme in which they pay a reduced rate of interest to the building society but cannot claim relief. (The building society claim the lost interest from the tax authorities.) What is their total tax allowance?

c) What is their taxable income?

d) If tax is charged at 25% on the first £750 taxable income and at 30% on the balance, what is the tax due *i*) per annum, *ii*) per week (to the nearest 10p)?

＊27 Where a worker's earnings are suddenly reduced, a refund of tax may be due. Consider the case where a man is earning £70 a week. His allowances total £1620 p.a.

a) What is *i*) his gross annual income, *ii*) his taxable income, *iii*) his tax at 30% per annum (give the tax for a full year and also for a week, to the nearest 1p)?

b) He works for 12 weeks, earning a total of £840. In the thirteenth week he earns nothing, so his income is £840 for 13 weeks. What would it amount to in a year at that rate?

c) What would be the taxable income at this new rate?

d) What would be the actual tax due for 13 weeks (i.e. $\frac{1}{4}$ year)?

e) How much tax did he actually pay in the first 12 weeks?

f) How much refund should he get in the 13th week?

＊28 If the man in question 27 again earned nothing in the 14th week, how much refund should he get in that week?

Further Note on Income Tax Allowances

a) The previous questions give a simplified picture of a very complex subject.

b) Allowances for children have not been included as they are being phased out and replaced by a 'child benefit' paid directly to the mother.

c) Insurance premiums on endowment or accident policies do not qualify for an income tax allowance.

d) The employer does not have to make complicated calculations. He knows the employee's coding, his earnings to date and the tax paid to date. Income tax tables then tell him the tax to be deducted or refunded that week.

Stocks and Shares

(*Questions 29–37 are based on current stock and share prices from Joseph Sebag's list, and on information received by courtesy of Messrs Daffern and Stephenson, Stock and Share Brokers, Coventry.*)

29 I instruct a broker to buy 200 ordinary shares in a well-known domestic appliance company. The nominal value of the shares is 25p and they are quoted at £3·00. Commission is 1·5% of the purchase price with a minimum charge of £7. Stamp duty is 2% of the purchase price.

a) How much do the shares cost me altogether?

b) If the dividend for the first half year after purchase is 35%, and for the second half year 54%, what total dividend do I receive?

c) What percentage does this represent on my outlay?

30 Repeat question 29 for 100 £10 shares in a water undertaking. The shares are quoted at £6·50 and pay a dividend of 7% for a full year.

31 The £1 ordinary shares in a ship canal are quoted at £2·40. They pay a dividend of 24% on a full year. Ignoring commission and stamp duty, what percentage return does an investor get on his outlay?

32 *a*) £100 4% debenture shares in the same company are selling at £30. What percentage return would these shares yield to the investor (again ignoring commission and stamp duty)?

b) Try to find out the difference between debenture shares and ordinary shares and hence explain why in this company debentures are quoted well below par and ordinary shares well above par.

33 An insurance company buy £100 000 $4\frac{1}{2}\%$ debenture stock in a chemical manufacturing firm. The stock is quoted at £44. Commission is charged at 1·5% on the first £7000 purchase price, 1·25% on the next £3000, 1% on the next £15 000 and 0·75% on the next £25 000. Stamp duty is 2% of the purchase price.

a) How much does the stock cost altogether?

b) What dividend does the insurance company receive in a full year, and what percentage is this on their total outlay?

34 An investor buys £5000 $5\frac{1}{2}\%$ debenture stock quoted at £80. Commission is charged at 1·5% on the purchase price and stamp duty at 2%.

a) How much does he pay altogether?

b) He receives half a year's dividend. How much is this?

c) He sells the stock at £81, and pays commission at $1\frac{1}{2}\%$ on the sale price. How much does he receive?

d) What is his net gain or loss over the six month period and what percentage does this represent on his original outlay?

35 Mrs Fortune's father dies and leaves her £20 000 $4\frac{1}{2}\%$ preference stock in a papermaking firm.

a) What is the annual income from this stock, if the full $4\frac{1}{2}\%$ is paid?

b) She sells it at £81, paying a total commission of £204·50. How much should she receive from the sale?

c) If she now instructs her broker to buy 9000 £1 ordinary shares in the same firm, these shares being quoted at £1·70, and if the commission on the sale is £193·50 and stamp duty is 2% of the purchase price, what do the shares actually cost her, and what balance should she receive from her broker?

d) If the shares pay a dividend of 12% for a full year, what annual income do they bring her?

e) Is she better off or worse off than before?

36 A retired lady teacher holds £1000 $3\frac{1}{2}\%$ debenture stock in a shipping company. She sells this stock at £25 and pays commission at the rate of 1·5% on the proceeds (minimum commission £7). She then buys 200 £1 preferred shares in the same company at 95p a share, and pays commission at 1·5% (with a minimum payment of £7) and £4 stamp duty.

a) At the end of the double transaction, what payment will she receive from or make to her stockbroker?

b) If the shares pay a dividend of 9·9% for a full year, has she raised or lowered her income? By how much?

37 An investor buys 500 £1 shares in a tea holding at £6·20 a share. Commission is 1·5% and stamp duty 2%, both calculated on the purchase price.

a) What is the total cost of his purchase?

b) If the dividend is 75% for the first full year and 85% for the second full year, how much dividend does he receive altogether in the two years?

c) In the following year, disease attacks the tea plants and the shares slump. He sells his holding at £4·20 a share, paying commission at 1·5% on the sale price. How much does he receive from the sale?

d) What is his total profit or loss, and what percentage is this of the original investment?

23B   Enlarged Horizons

Note   Some of the questions in this section will take you a little beyond the requirements of the current 'O' level syllabuses.

1   Can you see the fallacies in the following?

a) If $x=y=-4$, then by direct substitution $x^2+x-y^2-15y=56$;
rearranging, $x^2+x-12=y^2+15y+44$;
factorising, $(x+4)(x-3)=(y+4)(y+11)$.
But $x=y$ so $x+4=y+4$. Therefore $x-3=y+11$. But $x=y$, so $-3=+11$.

b) If $p=s=1$ and $q=r=-1$, then $ps=qr$, i.e. $\dfrac{p}{q}=\dfrac{r}{s}$ or $p:q=r:s$.

But $p>q$ so $r>s$, i.e. $-1>1$.

2   After a friendly rugger match between the Rhinos and the Armadillos, the two teams were presented with a bottle of rare Chateau Margaux. They decided to hold a draw, and Tom, the Armadillo manager, suggested lining up the thirty players in a circle and counting out every tenth man until only one man was left—he should have the bottle. George, the Rhino manager, agreed. Tom arranged the players and George did the counting, working clockwise. To his surprise the first fourteen players to be counted out were all Rhinos. He realised that Tom (who was a great practical joker) had been up to his tricks, so he suggested that the rest of the count should be anti-clockwise starting from the sole remaining Rhino. Tom agreed.

a) Calling the players A and R, how were they originally arranged?
b) Who won the bottle?

(*Sequel*   By mutual consent a fair draw was later held with thirty names in a hat.)

3   Here are some famous conjectures about the value of π:

Egyptians	3
Archimedes	between $3\frac{1}{7}$ and $3\frac{10}{71}$
Others	$\dfrac{355}{113}$

a) Taking the value of π as 3·14159, find the error in each expressed as a percentage of the true value of π.

b) If the value of π were known to 10 decimal places only, what would be the maximum error in calculating the circumference of a sphere of radius 6400 km, i.e. a sphere about the same size as the earth? Give your answer in millimetres.

4   Here are some interesting series for calculating π.

a) $\dfrac{\pi}{4}=1-\dfrac{1}{3}+\dfrac{1}{5}-\dfrac{1}{7}+\ldots$.

b) $\dfrac{\pi}{2}=1+\dfrac{1}{3}\left(\dfrac{1}{2}\right)+\dfrac{1}{5}\left(\dfrac{1.3}{2.4}\right)+\dfrac{1}{7}\left(\dfrac{1.3.5}{2.4.6}\right)+\ldots$

c) $\dfrac{\pi}{2}=\dfrac{2.2}{1.3}\times\dfrac{4.4}{3.5}\times\dfrac{6.6}{5.7}\times\ldots$

d) $\pi = 16\left(\dfrac{1}{5} - \dfrac{1}{3} \cdot \dfrac{1}{5^3} + \dfrac{1}{5} \cdot \dfrac{1}{5^5} - \dfrac{1}{7} \cdot \dfrac{1}{5^7} + \cdots\right) - 4\left(\dfrac{1}{239} - \dfrac{1}{3} \cdot \dfrac{1}{239^3} + \dfrac{1}{5} \cdot \dfrac{1}{239^5} - \cdots\right)$

Calculate the first to 12 terms, the next two to 8 terms and the last to 5 terms, taking the second bracket to 3 terms only. Give your answers to 5 d.p. Which series converges the most quickly?

5 In Book 1, 16B contained the following question (re-worded here):
Write down a three-figure number, the first digit being greater than the third. Reverse the order of the digits and subtract the new 3-figure number from the original. Retain any zeros that occur in any position. Reverse the digits again and add.
The rest of the question told you to multiply the result by 30 and decode, using the keyword PEWTER IRON to represent the digits 0 to 9. The result was always TWIRP.
You are now in a position to explain this algebraically. The key is that the number at the end of the addition always comes to 1089. Copy and complete the following table which proves that this is so. Start all additions and subtractions from the right. Remember that a is greater than c. Sometimes you will have to 'borrow ten' or 'carry one'.

Step	Operation	1st digit	2nd digit	3rd digit
1	start	a	b	c
2	reverse digits	c	b	a
3	subtract		9	$10 + c - a$
4	reverse digits			
5	add			

Pythagoras Triples

6 In Book 2, Chapter 14 there was a section on the calculation of Pythagoras triples. Here is another method of calculating these triples, based on the identity
$(a^2 + b^2)^2 - (a^2 - b^2)^2 = (2ab)^2$.
Copy and complete the table. This will give you all the 'prime triples' with no number larger than 50. It gives some but not all of the composite triples. (A prime triple p, q, r is one in which p, q and r have no common factor, e.g. 5, 12, 13. A composite triple is one in which two or more of p, q and r have a common factor, e.g. 6, 8, 10.)

a^2	b^2	$2ab$	p	q	r
			$(a^2 + b^2)$	$(a^2 - b^2)$	$(2ab)$
4	1	4	5	3	4
9	1	6	10	8	6
...	...	...			
49	1	14			
9	4	12			
...	...	...			
36	4				
16	9				
...	...	...			
25	16				

7 Here is yet another way of finding Pythagoras triples, based this time on the identity

$$\left(\frac{b+a}{2}\right)^2 - \left(\frac{b-a}{2}\right)^2 = (\sqrt{ab})^2.$$

This method gives all the triples, both prime and composite, and the table is laid out for the calculation of all triples with no number larger than 50.

a	b	p $\left(\dfrac{b+a}{2}\right)$	q $\left(\dfrac{b-a}{2}\right)$	r $\sqrt{ab}$
1	9	5	4	3
1	25			
...	...			
1	81			
2	8			
2	18			
...	...			
2	98			
3	27			
...	...			
...	...			
36	64	50	14	48

In compiling the table remember that a and b must be both even or both odd (as you have to divide their sum by 2), that ab must be a perfect square and that $a+b$ must not exceed 100.

Determinants, Equations and Matrices

8 The determinant shown on the right is a 3×3 determinant.

$$\begin{vmatrix} 3 & 1 & 2 \\ 2 & 2 & 3 \\ 1 & 0 & 2 \end{vmatrix}$$

a) Look at the number 3 in the first row and the first column. Knock out the first row and the first column.

You are left with the determinant $\begin{vmatrix} 2 & 3 \\ 0 & 2 \end{vmatrix}$. Its value is 4. This is called the 'minor' of the element 3. The minor of the element 2 in the second column and the second row is $\begin{vmatrix} 3 & 2 \\ 1 & 2 \end{vmatrix}$. Its value is 4. What is the minor of *i*) the element 3 in the third column and the second row, *ii*) the element 0 in the middle column of the bottom row?

b) The co-factor is the 'signed minor' of an element, i.e. it is the minor multiplied by $+1$ or -1. Count the number of steps by any *rectangular* path from the pivot to the element. If it is odd, the minor must be multiplied by -1. If it is even, multiply by $+1$, i.e. leave it unchanged. Thus the element 1 in the top row is one step from the pivot, so the co-factor is $-1 \times$ the minor. The element 2 in the bottom row is 4 rectangular steps from the pivot, so the co-factor is the same as the minor. What are the co-factors of *i*) element 2 in the second row and second column, *ii*) element 3 in the second row?

c) To evaluate the determinant, take the elements of any single row or column, multiply each in turn by its co-factor and add the results. Thus expanding the given determinant by the elements of the top row gives $3(4)+1(-1)+2(-2)=7$. Expand the determinant *i*) by the elements of the second row, *ii*) by the elements of the third column. In each case you should get the same answer, i.e. 7.

9 By expanding in terms of the elements of the top row, find the value of each of the following determinants:

a) $\begin{vmatrix} 3 & 2 & 1 \\ 2 & 1 & 0 \\ 1 & 4 & 1 \end{vmatrix}$
   b) $\begin{vmatrix} 2 & -1 & 1 \\ -2 & 1 & 2 \\ 1 & 3 & 1 \end{vmatrix}$
   c) $\begin{vmatrix} 1 & 0 & 1 \\ 1 & 1 & -2 \\ 0 & 1 & 1 \end{vmatrix}$

10 Check your answers to question 9 by expanding each determinant by the elements of the third column.

11 Expand the determinant shown. Multiply each element of the top row by its co-factor and add the results. The minors this time are 3×3 determinants. Check your answer by expanding in terms of the second row.

$\begin{vmatrix} 1 & 2 & 1 & 1 \\ 1 & 0 & 0 & 1 \\ 2 & 1 & 1 & 1 \\ 1 & -1 & 0 & 1 \end{vmatrix}$

12 Determinants give a quick way of writing down the area of a triangle when the co-ordinates of its vertices are known. Thus the area of the triangle whose vertices are (2, 2), (5, 3) and (4, 6) is given by the determinant shown. This is obtained by writing the co-ordinates as rows and adding a 1 at the end of each row.
Evaluate this determinant and find the area of the triangle.

$\frac{1}{2}\begin{vmatrix} 2 & 2 & 1 \\ 5 & 3 & 1 \\ 4 & 6 & 1 \end{vmatrix}$

13 a) Calculate the areas of the following triangles whose vertices are given:
   i) (1, 1), (5, 2), (3, 4)   ii) (4, 1), (1, 4), (5, 5)
   iii) (8, 6), (1, 5), (4, 1)   iv) (2, 2), (−1, 3), (1, −4)
   b) By splitting it into two triangles, find the area of the quadrilateral whose vertices are at (1, 2), (4, 3), (5, 7), (2, 4). Be careful to go round both triangles in the same direction (clockwise or anti-clockwise) or you will get one positive answer and one negative.

＊14 The determinant expression for the area of a triangle is easy, but tedious, to derive.
   a) In the figure, the area of trapezium $ABED$ is $\frac{1}{2}(y_1 + y_2)(x_2 - x_1)$. Write similar expressions for trapeziums $BCFE$ and $ACFD$.
   b) Triangle ABC is equal to $ABED + BCFE - ACFD$. Multiply out the three expressions you had in a, put them together and cancel any terms that can be cancelled. You should be left with $\frac{1}{2}(y_1x_2 - y_2x_1 + y_2x_3 - y_3x_2 + y_3x_1 - y_1x_3)$.
   c) Expand the determinant shown. You should get the same answer (with the sign changed).

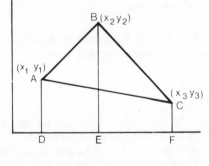

$\frac{1}{2}\begin{vmatrix} x_1 & y_1 & 1 \\ x_2 & y_2 & 1 \\ x_3 & y_3 & 1 \end{vmatrix}$

15 $3x + 2y + z = 6$
   $3x + y + 2z = 3$
   $x - y + z = -2$

To solve 'by elimination', multiply the second equation by 2 and subtract the first equation from the result. This gives an equation in x and z only. Now multiply the third equation by 2 and add the first equation to the result. This gives another equation in x and z only. Solve these two equations and find the values of x and z in the usual way. Substitution in any one of the original equations then gives y. Check your results by substituting in *all three* of the original equations.

16 Solve the following simultaneous equations:

a) $3x+2y+z=9$
$2x+y+2z=7$
$x-y+3z=4$

b) $x-y-z=4$
$2x+3y+z=10$
$3x+2y-2z=16$

c) $x+y-2z=-5$
$2x-y-z=-1$
$4x+3y+z=7$

*17 Try to solve the following equations:

a) $2x+3y-z=3$
$5x+4y+z=4$
$x-2y+3z=-2$

b) $3x-2y+z=9$
$5x+y-z=12$
$2x+3y-2z=-1$ What happens?

*18 The equations in 17a are *redundant*. Any one equation is a *linear combination* of the other two. Thus the second equation is 'the third plus twice the first', and the third equation is 'the second minus twice the first'. There are therefore effectively only two equations, and an infinity of possible solutions.

a) To find a set of solutions, eliminate z from the first two equations. This gives $x+y=1$. Possible solutions are $x=4$, $y=-3$ or $x=2$, $y=-1$ or $x=-1$, $y=2$, etc. For each of these find the value of z (using the first equation).
b) You now have three sets of solutions. Find two more.

19 The equations in 17b are rather different from those in 17a. They are incompatible. If you add the first and the last you get $5x+y-z=8$, whereas the second equation is $5x+y-z=12$. These results cannot both be true at the same time, so there is no solution to the equations.

a) What could you do to the first equation in 17a to make that set incompatible?
b) Find the value of the determinant of the coefficients in each of 17a and 17b. What do you notice?

20 Solve the equations in question 16 using Crout's method (see Section 6D). Is there any advantage over elimination?

21 Find the inverse of the given matrix by following the procedure stated:

$$\begin{pmatrix} 1 & 2 & 1 \\ 2 & 3 & 4 \\ 1 & -2 & 1 \end{pmatrix}$$

a) Make a fresh matrix by writing down the co-factor of each element instead of the original element.
b) Transpose this new matrix, i.e. write its rows as columns and its columns as rows, keeping the pivot unmoved.
c) Divide the new matrix by the determinant of the original matrix. Call the result B.
d) Calling the original matrix A, check that $AB=BA=I$, i.e. that B is the inverse of A.

22 Invert the following matrices, checking in each case that you have the true inverse (as in 21d).

a) $\begin{pmatrix} 3 & 1 & 2 \\ 4 & -2 & 1 \\ 1 & 0 & -1 \end{pmatrix}$

b) $\begin{pmatrix} 2 & 6 & 1 \\ 1 & -1 & 0 \\ 0 & 2 & 1 \end{pmatrix}$

c) $\begin{pmatrix} 2 & -1 & -2 \\ 3 & 0 & 1 \\ 1 & 1 & 0 \end{pmatrix}$

*23 Solve the equations in question 16 by matrix inversion.

Magic Squares

✱✱24 In Book 2 a recipe was given for making magic squares, i.e. square arrays of numbers in which the rows, columns and diagonals all have the same sum. You now know enough about algebra to be able to derive this recipe for yourself.

a) Here is the recipe. Choose any values you like for a and i. e is then $\frac{1}{2}(a+i)$. p, the sum of the rows, etc., is $a+e+i$ and will always be a multiple of 3. Now choose g. It must be less than $a+e$, less than $i+e$ and more than half the difference between a and i. Copy and complete the square shown, choosing your own value for g.

a	b	c
d	e	f
g	h	i

6		
		16

b) Make up another square in which g lies outside the limits laid down. You will find you get negatives.

c) To investigate the recipe, proceed as follows.
There are 8 equations connecting the letters $a, b, \ldots i$ and the sum of the rows, etc. (which we will call p). Here are the eight equations:

i) $a+b+c=p$ ii) $d+e+f=p$ iii) $g+h+i=p$ iv) $a+d+g=p$
v) $b+e+h=p$ vi) $c+f+i=p$ vii) $a+e+i=p$ viii) $c+e+g=p$

One of these eight equations is redundant. You can see this easily by adding i), ii) and iii). It gives the same result as adding iv), v) and vi) so there are effectively seven equations and ten variables (including p). This means we can choose values for three variables, and the rest are then fixed. We will choose a, i and g. There are, however, restrictions on their values or we shall end up with negatives and fractions.
First restriction: a and i must both be even or both odd, e must be $\frac{1}{2}(a+i)$ and p must be a multiple of 3. To show this, add equations i) and iii) and subtract v) and viii) from the result. What do you get? Does this prove the first restriction?
Second restriction: g must be less than $a+e$, less than $i+e$ and greater than half the difference between a and i. To show this, express all the other letters in terms of a, e, i and g. Replace p by $a+e+i$, so that $p-a$ is $e+i$, $p-i$ is $e+a$. Start with viii) which gives $c=a+i-g$. iii) and iv) give similar expressions for h and d. Finally i) and ii), using the values of c and d already calculated, give the values of b and f. Can you deduce the second restriction from these five results?

✱25 If you want all the terms in a magic square to be in arithmetic progression with constant difference q, then the rules still hold but there are additional restrictions.
For example, choose 3 for the common difference q. i must be greater than q. Let $i=5$. a must be $6q$ greater than i, i.e. 23. e is $\frac{1}{2}(a+i)$ as before, i.e. 14, and g must be $e-q$, i.e. 11. The rest follows.

23		
	14	
11		5

a) Complete the square. Check that the terms form an arithmetic progression.
b) Make up other squares with i) $i=7, q=4$ ii) $i=2, q=1$ iii) $i=5, q=2$.

Miscellaneous Examples C

C1

1 Find the solution set for the equation $(x+2)(2x-1)(x^2-5)=0$ in the set of:
a) positive integers   b) rational numbers   c) real numbers
d) irrational numbers.
Leave your answers to c and d as surds.

2 If $p*q$ denotes $(p+q)^2$ for any rational numbers p and q, evaluate:
a) $2*3$ b) $-4*3$ c) $1*(4*2)$.
d) If $q=2p$ and $p*q=144$, find p and q.
e) If $p*q=p^2$ and $q\neq0$, express p in terms of q.

3 $(0, 0)$, $(2, -2)$, $(4, 0)$ and $(2, 2)$ are the co-ordinates of the vertices of a
quadrilateral. This quadrilateral is transformed by the matrix $\begin{pmatrix} 1 & -1 \\ 1 & 1 \end{pmatrix}$. Find
a) the positions of the vertices of the transformed figure,
b) its area,   c) the area of the original quadrilateral.

4 Two of the sides of a right-angled triangle are 10 cm and 24 cm respectively. Find
a) the possible lengths of the third side of the triangle,
b) the possible values of the area of the triangle.

5 Find the value of $\dfrac{\sqrt{360 \times 29}}{11\cdot7}$ correct to 2 significant figures.
Write down the value of $\dfrac{\sqrt{36\,000 \times 2900}}{1\cdot17}$ in standard form.

6 The probability that I stop to buy a newspaper on my way to work is $\frac{3}{5}$ and the probability that I stop to post a letter is $\frac{2}{9}$. The post-box is not outside the newsagents so I sometimes stop twice. Find the probability that

a) I do not stop to post a letter
b) I stop to post a letter and buy a paper
c) I do not stop at all.

7 Given $f: x \to 8-3x$ find   a) $f(3)$ b) $ff(1)$
c) a if $f(a)=3\cdot5$ d) $f^{-1}(x)$.

8 Find the gradient of the line joining the points $(0, -2)$ and $(2, 1)$ and hence write down the equation of this line.

9 The marks gained in a test by ten pupils were: 9, 17, 15, 14, 11, 14, 18, 12, 13, 12.

a) Find the median of these marks.   b) Find the mean.
c) Two pupils took the test late and the mean decreased to 13·3. One scored 15. What did the other one score?

10 Match the equations below with the graphs.

$$y=\sqrt{x} \qquad y=1/x \qquad y=1/x^2 \qquad y=x^2 \qquad y=x^3 \qquad y=2x \qquad y=2^x$$

a) 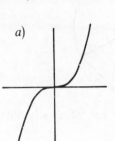 b) 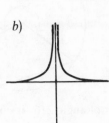 c) d)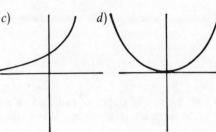

e) 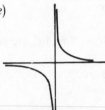 f) 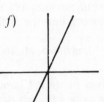 g)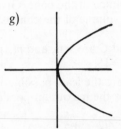

11 The interior angle of a regular polygon is four times the size of the exterior angle. How many sides does the polygon have?

12 Place one of the signs $\Rightarrow$ or $\Leftrightarrow$ or $\Leftarrow$, where appropriate, between the following statements. If none is appropriate, say so.

a) $\sin A = \cos B \qquad A+B=90°$.
b) $ABCD$ is a parallelogram $\qquad ABCD$ has symmetry about AC.
c) a and b have no common factor $\qquad a$ and b are different prime numbers.
d) $P \cup Q = P \qquad Q \subset P$.
e) angle $ABC = 90° \qquad AC$ is a diameter of a circle which passes through B.
f) $x=4 \qquad x^2-x-12=0$.

C2

1 The line $y=2x$ is reflected in the line $y=0$. What is the equation of its image? What is the equation of the image of $y=2x$ after reflection in the line $y=2$?

2 If $p \propto q^2$ and $p \propto \dfrac{1}{r}$, write an equation for r in terms of q given

that $p=3$ when $q=2$ and $r=-1$. Which value is superfluous?

3 If angle A is less than $180°$ and $\tan A = -\dfrac{5}{12}$, write down the values

of $\sin A$ and $\cos A$.

4 In the figure, O is the centre of the circle. Write down the angles x and y in terms of a and b.

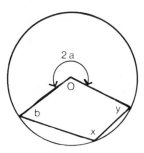

5 From a circle of paper of radius R, a sector of angle $216°$ is cut and folded without overlap to form a circular cone. Write down:

a) the radius of the base of the cone   b) the area of the sector

✳ c) the volume of the cone. Give your answers in terms of R and/or π.

d) If the volume of the cone is 96π cm, find the radius R.

6 A, B and C are sets, and $n(A)=13$, $n(B)=5$, $n(C)=10$. If A, B and C are subsets of $\mathscr{E}$ where $n(\mathscr{E})=18$,

a) what are the least possible values of $n(A\cap C)$ and $n(B\cup C)$?

b) what is the maximum possible value of $n(A\cap B\cap C)$?

7 a) Solve $4(2-x)<3(5-x)$.   b) Find x and y if $\begin{pmatrix} -2 & 1 \\ 3 & 0 \end{pmatrix}\begin{pmatrix} x \\ y \end{pmatrix}=\begin{pmatrix} 4 \\ -3 \end{pmatrix}$.

8 The length and breadth of a rectangle are measured to the nearest cm and found to be 21 cm and 15 cm respectively. Find the largest and smallest values of the area consistent with these measurements, and write your answers in the form $A_1\leqslant \text{area}<A_2$.

9 P is the point $(2, 1)$. M represents reflection in the line $x=5$.

T represents translation by the vector $\begin{pmatrix} -4 \\ 3 \end{pmatrix}$.

Write down the following images of P:

a) $M(P)$ b) $T(P)$ c) $TM(P)$ d) $M^2(P)$.

10 Given that n is a positive integer and $G(n)$ is the set of factors of n including 1 and n, find: a) $G(12)$ b) $G(18)$ c) $G(30)$ d) $G(12)\cap G(18)\cap G(30)$.

Describe in words the meaning of your answer to d.

11 Two identical cubes are coloured in three colours, red, green and blue, with opposite faces the same colour. If the two are tossed together, what is the probability of a) both showing red,   b) both showing the same colour?

If a third similar cube is thrown with the other two, what is the probability of c) all showing red,   d) each of the three showing a different colour?

12 A car travels for 20 minutes at an average speed of 48 km/h and a further 16 km at an average speed of 40 km/h. Calculate the average speed for the whole journey.

C3

1 The curve $y=ax^2+bx+c$ passes through the points $(0, -2)$, $(1, 4)$ and $(-1, -6)$. Find the values of a, b and c and the equation of the curve.

2 If $4x^2-12xy+9y^2=0$, find the value of x/y.

3 In the diagram, $BC=6$ cm and angle $C=90°$.
90°.

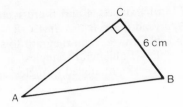

a) If angle $A<30°$, write down the range of values for AB.

b) If angle $B<45°$, write down the range of values for AC.

4 a) If $u=\binom{2}{3}$, $v=\binom{-1}{2}$ and $w=\binom{4}{-1}$, find the value of k if $u+kv=w$.

b) Translations A and B are defined by the vectors $\binom{3}{1}$ and $\binom{2}{4}$ respectively.

The image of the point $P(1, 2)$ after translation A is Q, and the image of P after translation B is R. Write down the co-ordinates of Q and R and prove that PQ is perpendicular to QR.

5 In the diagram, BC is parallel to DE, $AB=3$ cm, $BD=2$ cm and $BC=4·5$ cm. Calculate the length DE. If the area of ABC is $4·5$ cm^2, find the area of $BCED$.

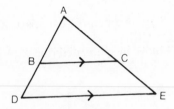

6 A triangle T is formed by joining the points $(1, 1)$, $(2, 1)$ and $(1, 3)$. A denotes a rotation through $90°$ clockwise about $(3, 1)$. B denotes a reflection in the line $y=1$. Draw a diagram to show $A(T)$ and $BA(T)$. Find a single transformation which maps T on to $BA(T)$.

7 a) Factorise i) $9x^2-y^2$ ii) $16x^4-a^4$.
b) Calculate the exact value of $7·64^2-2·36^2$.

8 A bill from the decorator came to £121·50 including 8% VAT. What was the total excluding VAT?

9 The latitude and longitude of Johannesburg are $26°$ S and $28°$ E. Write down the latitude and longitude of the place which is at the other end of the diameter

a) of the circle of latitude through Johannesburg,
b) of the circle of longitude through Johannesburg.

10 A circle is divided into three sectors A, B, C so that the angles at the centre are in the ratio $5:3:1$. The area of sector A is 16 cm^2 larger than the area of sector B. Calculate

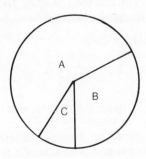

a) the angle of sector C,
b) the area of the circle,
c) the radius of the circle.

11 Find the area of a triangle with sides of 12 cm, 9 cm and 15 cm.

12 *a*) Two sets A and B are such that $A \cap B \neq \phi$ and $A \cup B = \mathscr{E}$.
Find *i*) $A \cap B'$ *ii*) $A' \cup B$.
b) Draw a Venn diagram to show these sets:
$E = \{$Equilateral triangles$\}$ $I = \{$Isosceles triangles$\}$
$R = \{$Right-angled triangles$\}$ $O = \{$Obtuse-angled triangles$\}$

C4

1 *a*) Find the solution set if $(x-5)(x+2) < 0$.
b) Find the least value of $(x+2)^2 + 3$.

2 The lines $y = x$, $y = x+3$, $x+y = 3$ and $x+y = 8$ intersect at four points and enclose a rectangle. Write down the lengths of the sides of the rectangle in surd form and hence find its area.

3 A and B are two points 8 cm apart. $X = \{P$: angle $APB = 90°\}$ and $Y = \{Q : AQ = QB\}$.
If P and Q move in space, describe X, Y and $X \cap Y$.

4

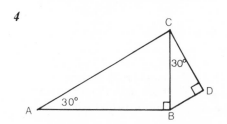 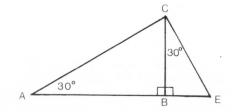

In each of the diagrams, AC is of length $4a$ units. Write down:

a) BC *b*) BD *c*) BE *d*) $\dfrac{\text{area of triangle } ABC}{\text{area of triangle } CDB}$

e) $\dfrac{\text{area of triangle } ABC}{\text{area of triangle } CBE}$ *f*) $\dfrac{\text{area of triangle } CDB}{\text{area of triangle } CBE}$

5 If $\overline{OA} = a$ and $\overline{OB} = b$, and M is the mid-point of AB, find $\overline{OM}$ in terms of a and b.

6 A pyramid has a square base of side 6 cm and is 7 cm high. The vertex P is vertically above the centre of the base $ABCD$.
Calculate the angles between

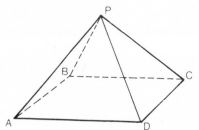

a) AP and the base,
b) the face PCD and the base,
c) the faces PAB and PCD.

7 *a*) Two ships on the equator are 1950 nautical miles apart (measured round the equator). Calculate the difference in their longitudes.
b) Two points on the circle of latitude 60° N are 1950 nautical miles apart (measured round the circle of latitude). Calculate the difference in their longitudes.
c) Repeat *a* and *b* if the distances are 3000 km, given that the radius of the earth is 6366 km.

8 Given $f : x \rightarrow (x+3)(6x^2 - 5x + k)$ and that $f(2) = 40$, find the value of k. Hence calculate the values of x for which $f(x) = 0$.

9 The cross-section of a metal rod is an isosceles right-angled triangle, with the equal sides measuring 4 cm exactly. A short length is cut off. If this measures 2·6 cm to the nearest millimetre, find the greatest possible error in calculating the volume of this prism. Express this error as a percentage of the calculated volume.

10 Simplify: *a)* $(3\sqrt{2} - 2)(5 + 2\sqrt{2})$ *b)* $(2\sqrt{3} - \sqrt{2})(2\sqrt{3} + \sqrt{2})$

     c) $\dfrac{2}{3\sqrt{2} - 3} \times \dfrac{1 + \sqrt{3}}{\sqrt{2} + 1}$

11 *a)* If $\sin A = \cos B = -\frac{1}{2}$, find angles A and B, both of which lie between 180° and 270°.
   b) Angles C and D add together to make 90°. If $\tan C = \frac{5}{7}$, what is $\tan D$?
   c) Angles E and F add together to make 180°. If $\tan E = \frac{1}{3}$, what is $\tan F$?

12 In the diagram, O is the centre of circle ABC and BT is a tangent.
If angle $AOB = 2a$ and angle $OBC = b$, write down angle OAC in terms of a and b.

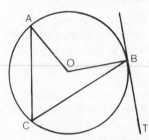

C5

1 The lines $3x - 2y = p$ and $2x + qy = 3$ intersect at the point $(4, -1)$. Find the values of p and q.

2 $ABCDEF$ is a regular hexagon, whose centre is O.

   a) State the image of triangle ABO
   i) after reflection in BE,
   ii) after rotation about O through 60°.
   b) State the transformation which maps
   i) triangle AFO on to triangle EDO,
   ii) triangle BCO on to triangle EFO.

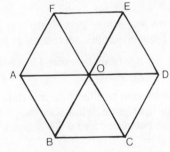

3 If $A = \{x : -2 \leqslant x < 3\}$, $B = \{x : 1 < x \leqslant 5\}$ and $C = A \cap B'$, find C.

4 The ratio of the surface areas of two spheres is $4:25$. If the volume of the smaller sphere is 64 cm³, what is the volume of the larger one?

5 Simplify: *a)* $\sqrt{32} + \sqrt{8}$ *b)* $\sqrt{32} \times \sqrt{8}$ *c)* $\sqrt{2^3 \times 3^2}$ *d)* $\sqrt{\dfrac{3 \cdot 2 \times 10^9}{5 \cdot 0 \times 10^5}}$

6 If $\sin A = 0 \cdot 8$ and $0 \leqslant A \leqslant 180°$, find the values of $\cos^2 A$ and $\tan A$.

7 Functions f and g are defined by $f : x \rightarrow x^2$ and $g : x \rightarrow x + 2$.
Write down the functions $fg(x)$ and $gf(x)$. Find the range of values of x for which $fg(x) < gf(x)$.

8 Find the volume and total external surface area of the bread bin shown in the diagram. (Take π as $\frac{22}{7}$.)

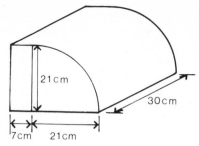

9 $p*q$ denotes the highest positive integer which is a factor of both p and q.
a) Find the value of: *i)* $10*14$ *ii)* $18*30$ *iii)* $8*28$.
b) If $m*n=n$, what can you say about m and n?
c) If $x*y=1$, what can you say about x and y?

10 AB is the tangent at A to a circle whose centre is O. OB cuts the circle at C. $AB=12$ cm and $BC=8$ cm.

a) Calculate the radius of the circle.
b) A second tangent is drawn from B to meet the circle at D. Prove that a second circle can be drawn to pass through O, A, B and D.
c) What is the radius of this circle?
d) Calculate also the length of the common chord AD.

11 $P=\begin{pmatrix} 1 & 0 \\ 5 & a \end{pmatrix}$ and $Q=\begin{pmatrix} 2 & 0 \\ -1 & 3 \end{pmatrix}$.
Calculate *a)* PQ, *b)* QP, *c)* the value of a when $PQ=QP$, *d)* Q^{-1}, *e)* the value of a for which P^{-1} does not exist.

12 In a survey, people going into a supermarket were stopped and asked their ages. The first 200 who replied gave ages as follows:

Age (years)	under 15	15–20	20–40	40–65	65–75	over 75
Frequency	0	14	96	75	15	0

Draw a histogram to illustrate this data.
From your histogram, what was the probability that the first person who answered was *a)* aged between 15 and 20, *b)* aged between 25 and 30, *c)* aged 64?

C6

1 $PQRS$ is a trapezium in which PQ is parallel to SR. If $PQ=9$ cm, $QR=8$ cm and $PS=PR=7$ cm, calculate the length of SR. Find also the area of the trapezium.

2 During the summer sale two shops, X and Y, are selling clothes at the following prices (in pounds):

	Shop X	Shop Y
Dresses	8	9
Skirts	3	2
Blouses	2	2

This matrix is denoted by A.

Gill and Betty go separately on a shopping spree. Gill buys only at Shop X, and Betty only at Shop Y. The following matrix represents their purchases:

	Dresses	Skirts	Blouses
Gill (Shop X)	1	1	2
Betty (Shop Y)	1	3	1

This matrix is denoted by B.

a) Calculate the matrix product $C = BA$. Describe carefully in words what is represented by i) the first number in the top row of C ii) the second number in the top row of C.

b) Calculate the matrix product $D = AB$. Describe carefully in words what is represented by the first number in the top row of D. (MEI)

3 O is the origin and the points A and B have position vectors $\overline{OA} = \boldsymbol{a}$ and $\overline{OB} = \boldsymbol{b}$. L is the mid-point of OB, M is the mid-point of AB and N the mid-point of AM. The lines ON and LM produced meet at P. Find the position vectors of L, M and N and prove that $\overline{OP} = \frac{3}{2}\boldsymbol{a} + \frac{1}{2}\boldsymbol{b}$. Hence or otherwise show that BP is parallel to LN and twice its length. (OXFORD)

4 The gradient of a curve which passes through the origin is given by $dy/dx = 3x^2 - 6x + 2$.

a) Find the equation of the curve.

b) Find the total area cut off by the x axis and the curve between $x = 0$ and $x = 2$. What can you deduce?

c) This curve and the curve $y = x^2 - 2x$ have the same gradient for two different values of x. Find these values of x.

5 My job is such that as well as Sunday off each week I have either Saturday or Tuesday. The probability of Saturday off is $\frac{1}{3}$, On my day off, I like to go to a football match played at home. The probability of there being a match on a Saturday is $\frac{3}{5}$, but the probability of there being a match on a Tuesday is only $\frac{1}{5}$. What is the probability of

a) having Tuesday off,

b) being able to go to a match on a Saturday,

c) being able to go to a match on my day off?

If the teams which I support have a chance of $\frac{1}{3}$ of winning at home, what is the probability of

d) seeing a match won on a Saturday,

e) seeing a match won on my day off?

6 A plantpot has a base radius of 6 cm and the radius of the top is 7·5 cm. Its height is 10 cm. The curved sides of the pot form the bottom part of the curved surface of an inverted cone.

a) Find the height of the whole cone. Calculate also the capacity of the plantpot and its total external surface area.

b) If the plantpot were made to rest on its curved surface, calculate the angle that the base would make with the horizontal.

C7

1 At 1400 hours, a ship is on a bearing of 140° from a lighthouse L. The ship is cruising at a speed of 30 km/h on a course 255°. At 15 00 hours it is at A, a point due south of the lighthouse. Calculate *a*) the distance *AL*, *b*) the time at which the ship is nearest to the lighthouse (to the nearest minute).

2

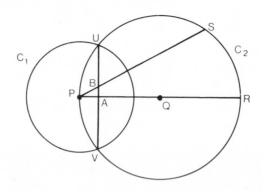

Two circles C_1 and C_2 intersect at U and V, and C_2 passes through P the centre of C_1. The point Q is the centre of C_2 and PQR is a diameter of C_2. The point S is an arbitrary point of C_2 lying on the minor arc between U and R. The chord UV interesects PQ at A and PS at B. Prove that

a) the triangles *PBA* and *PRS* are similar
b) $PB \cdot PS = PA \cdot PR$
c) triangles *PUA* and *PRU* are similar
d) $PU^2 = PA \cdot PR$
e) the two circles *UAR* and *UBS* touch at *U*.

(MEI)

3 In parallelogram *PQRS*, the diagonal *PR* is at least equal in length to the diagonal *QS*.

a) If angle *QPS* is $x°$ and angle *PQS* is $y°$, write down an expression for angle *PSQ*.
b) The length of *PS* is equal to or greater than the length of *PQ*, and *SQ* is equal to or greater than *PS*. Explain why each of the following inequalities is true: *i*) $x \leqslant 90°$, *ii*) $x + 2y \geqslant 180°$, *iii*) $x \geqslant y$.
c) On a graph, using a scale of 1 cm to 10° on each axis, draw the three lines $x = 90$, $x + 2y = 180$ and $x = y$, and shade the areas containing the pairs of values of x and y which are ruled out by the inequalities. From your graph *i*) find the smallest possible value of angle *QPS*, *ii*) find the greatest possible angle between *QS* and the shorter sides of the parallelogram, *iii*) name the type of parallelogram formed by *P, Q, R* and *S* when the difference between x and y is as large as possible.

4 Three matrices are defined as follows:

$$E = \begin{pmatrix} 1 & 0 \\ 0 & 1 \end{pmatrix} \quad F = \begin{pmatrix} -1 & 0 \\ 0 & -1 \end{pmatrix} \quad G = \begin{pmatrix} 1 & 2 \\ -1 & -1 \end{pmatrix}$$

a) Find a matrix H such that the set E, F, G, H is closed under matrix multiplication.
b) Copy and complete the table.
c) For the table in *b* write down
i) the identity element
ii) the inverse of *H*
iii) the inverse of *F*.

×	E	F	G	H
E	E	F	G	H
F	F			
G				
H	·			

(MEI)

218

5 A scientist assumes that three variable quantities x, y, z are connected by an equation $z = x(p + q\sqrt{y})$, where p and q are constants.

a) Derive the values of p and q from the results
i) when $x = 5.0$ and $y = 9.0$, then $z = 40.5$,
ii) when $x = 8.0$ and $y = 16.0$, then $z = 84.8$.
b) If the constants are (wrongly) taken as $p = 0.3$ and $q = 2.6$, what would be the calculated value of z for $x = 6.6$ and $y = 12.0$, and what would be (to one significant figure) the percentage error in this value compared with the value $z = 60.2$ obtained by experiment?
c) Restate the equation with y as subject, showing each step in your working.

(OXFORD)

6 During a traffic survey, on entering a car park in a busy city centre drivers were asked how long they had been driving that morning. The times stated by ninety drivers are shown in the table.

Time in minutes	1–10	11–20	21–30	31–40	41–50	51–60	61–70	71–80
No. of drivers	2	16	22	25·	11	6	4	4

Draw a cumulative frequency graph and from this or from the original table find answers to the following questions:

a) What is the mean length of time of a journey?
b) What is the median time?
c) What percentage of the drivers had driven for 15 minutes or less?
d) What percentage had driven for 45 minutes or longer?

C8

1 ABC is a triangle in which angle $A = 90°$, $AC = 2$ cm and $AB = 5$ cm. D is the reflection of A in BC. Calculate the radius of the circle which passes through A, B, D and C and the angle which AD subtends at the centre of the circle.

2 The matrix P records the numbers of each 'O' level grade obtained in mathematics by three forms at a certain school.

$$
P \quad \begin{array}{c} \\ \text{Form 1} \\ \text{Form 2} \\ \text{Form 3} \end{array} \begin{array}{c} \text{Grade} \\ \begin{array}{cccccc} A & B & C & D & E & U \end{array} \\ \begin{pmatrix} 6 & 5 & 8 & 5 & 2 & 0 \\ 2 & 4 & 5 & 6 & 2 & 1 \\ 0 & 2 & 3 & 8 & 4 & 2 \end{pmatrix} \end{array}
$$

a) Evaluate $P \begin{pmatrix} 1 \\ 1 \\ 1 \\ 0 \\ 0 \\ 0 \end{pmatrix}$ Explain the significance of the numbers in the matrix that you obtain.

b) It is required to find the numbers in each form obtaining a D or E grade. Write down a similar product of matrices to that in *a* which would give this information. (You are not asked to evaluate this product.)
c) Find a matrix Q so that the product PQ has three rows representing the three forms, and four columns, the first column being the number of pupils obtaining an A, B or C grade, the second the number obtaining a D or E grade, the third the number obtaining a U, and the fourth being the total number of pupils in that form.

d) Evaluate *PQ*.

e) Suggest a reason why $PQ \begin{pmatrix} 1 \\ 1 \\ 1 \\ -1 \end{pmatrix}$ might be evaluated.

f) Evaluate $(1 \quad 1 \quad 1) \, PQ$, and state what information is recorded in it. (MEI)

3 *a*) For the curve $y = x^3 + x^2 - 8x + 7$, find the values of x for which y is a maximum or a minimum. Find these values of y and say which is the maximum and which the minimum.

b) A particle moving in a straight line passes a point *P* when $t = 0$. At that moment it is moving with a velocity of 5 cm/s. At time *t* seconds, the distance of the particle from *P* is *s* cm, its velocity is *v* cm/s, and its acceleration is $(4t - 7)$ cm/s². Find *i*) an equation for *v* in terms of *t*, *ii*) the times at which the particle is momentarily at rest, *iii*) an equation for *s* in terms of *t*, *iv*) the distance of the particle from *P* when it is first at rest.

4 The points *A*, *B*, *C* lie on the circumference of a circle. Tangents to the circle at *A* and *B* meet at *X*. The letters *p*, *q*, *r* denote these three statements: *p* is 'XA and BC are parallel'; *q* is 'the triangle *ABC* is isosceles'; *r* is 'the triangles *BXA*, *ABC* are equiangular with angle *BXA* = angle *BAC* and angle *XAB* = angle *ACB*'. Prove *i*) $p \Rightarrow q$, *ii*) $p \Rightarrow r$. State with reasons whether *iii*) $q \Rightarrow p$, *iv*) $r \Rightarrow p$. (OXFORD)

5 In this question, $\mathscr{E} = \{$Senior pupils in a mixed school$\}$, $G = \{$Girls$\}$, $F = \{$Pupils who take French$\}$, $M = \{$Pupils who take mathematics$\}$, $P = \{$Pupils who take physics$\}$, $E = \{$Pupils who take economics$\}$, $W = \{$Pupils who do woodwork$\}$.

a) Express these sentences in symbols: *i*) All pupils take at least one of the subjects French, economics or woodwork. *ii*) Only boys do woodwork. *iii*) No pupil takes physics who does not also take mathematics.

b) Write sentences to express the following statements: *i*) $E \cap M' = \phi$, *ii*) $F \cap P = F$.

c) If all these five statements are true, then there is one subject which is taken by every girl. Name the subject and give reasons for your answer.

6 A garage owner decides to start a car hire business with two types of car. Type *A*, the smaller, costs £3500 each and type *B* costs £4500 each. The money available is £63 000 and the owner wants to buy at least 12 cars. He decides that the number of *A* must be at least twice the number of *B* and that there must be at least three of *B*. If he buys *x* of *A* and *y* of *B*, write down four inequalities in *x* and *y* and show them on a graph. Use 2 cm to represent two cars on each axis and show by shading the region in which *x* and *y* cannot lie. From your graph find how many of each type of car he must buy to make the total number of cars a maximum.

C9

1 The country of Northland has four domestic airports *E*, *F*, *G*, *H* and three international airports *L*, *M*, *N*. The matrix

$$A = \begin{array}{c} \\ E \\ F \\ G \\ H \end{array} \begin{array}{ccc} L & M & N \\ \begin{pmatrix} 1 & 1 & 0 \\ 1 & 0 & 1 \\ 0 & 1 & 0 \\ 1 & 0 & 0 \end{pmatrix} \end{array}$$

indicates whether or not there are flight connections between each domestic airport and each international airport. An entry of 1 in the matrix denotes that there is such a connection, and an entry of 0 denotes that there is not.

The country of Southland has two international airports, P and Q. The existence or otherwise of flights between the international airports of Northland and Southland is given by the entries in the matrix

$$B = \begin{array}{c} \\ L \\ M \\ N \end{array} \begin{array}{c} P \quad Q \\ \begin{pmatrix} 1 & 0 \\ 1 & 1 \\ 1 & 0 \end{pmatrix} \end{array}$$

Calculate the matrix product AB and interpret the entry in the first row and first column.

Southland also has two domestic airports X and Y. There are domestic flights to Y from both P and Q, but airport X can only be reached from P. Write this information as a matrix C in such a way that the matrix product $(AB)C$ will give the numbers of different routes from domestic airports in Northland to domestic airports in Southland. Evaluate this matrix product. (MEI)

2 a) OAB is a triangle and $\overline{OA} = a$, $\overline{OB} = b$. Draw a diagram showing OAB and mark on it the point P whose position vector $\overline{OP}$ is $\frac{1}{2}b$, the point Q whose position vector is $\frac{1}{2}(a+b)$ and the point G whose position vector is $\frac{1}{3}(a+b)$.
Use vector methods to prove that A, G and P are collinear and to find the ratio $AG:GP$.
b) x and y are two vectors which are not parallel. A third vector z can be expressed both as $px + qy$ and also as $y + q(x - y)$ where p and q are numbers. Find p and q. (OXFORD)

3 a) The formula $P = \dfrac{k}{\sqrt{1+t}}$ gives P in terms of k and t, where $k = 1.65 \times 10^2$ and t is a positive number. Make t the subject of this formula and use logarithms to calculate the value of t when $P = 1.32 \times 10^2$.
b) If t is small, an approximate formula for P is $P = k(1 - \frac{1}{2}t)$. Evaluate P when $t = 0.21$ both from the original and from this approximate formula. What is the percentage error in this case using the approximate formula? (MEI)

4 The figure shows a circle, centre O, of radius 2 units, and a point A on the circle. B is the other end of the diameter through A. The mediator (perpendicular bisector) of OA cuts the circle at X and Y, and the diameter AB at C.

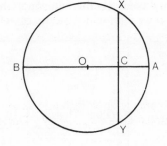

a) Copy the figure and draw the reflection of the circle in the line XY. Mark the image of B as B'. Write down the size of the angle OXA. Calculate angle OXB' and state what you can deduce about the lines $B'X$ and $B'Y$ and the circle $AXBY$.
b) The figure $OXB'Y$ is rotated through $120°$ clockwise about O, and also $120°$ anticlockwise about O. Under these transformations B' maps to P and Q respectively. Prove that PBQ is a straight line.

c) The figure $OXAY$ maps to $XB'YB$ under a rotation of 90° clockwise about C followed by an enlargement with centre C. Show that the scale factor of the enlargement is $\sqrt{3}$.

(OXFORD)

5 The curve $y = 5x - 4 - x^2$ cuts the x axis at A and B, A being nearer to the origin than B.

a) Find the x co-ordinates of A and B.
b) By differentiating, find the gradient of the curve at any point (x, y).
c) C is the point whose x co-ordinate is 2. Find: *i)* the y co-ordinate of C, *ii)* the gradient of the tangent to the curve at C, *iii)* the equation of the tangent to the curve at C.
d) Calculate the area enclosed by the curve, the tangent at C and the x axis.
e) Hence calculate the area enclosed by the curve, the tangent and the y axis.

6 Given that $f: x \to 2^x$ for $-3 \leqslant x \leqslant 4$, draw the graph of $f(x)$ using a scale of 2 cm to 1 unit on the x axis and 2 cm to 2 units on the other axis.
Draw the tangent to the curve at $x = 2$ and estimate the gradient of $f(x)$ when $x = 2$.
Find also the area between the curve, the x axis and the lines $x = 1$ and $x = 3$.

C10

1 The diagram shows two circles which meet each other at C. AB is a common tangent meeting the circles at A and B respectively. The common tangent at C meets AB at T. The chords AC and BC are produced to meet the other circles at E and D respectively. Prove that:
  a) AD is parallel to BE
  b) angle ACB is 90°.

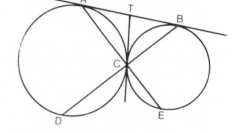

2 The cost (£C) of flying a freighter aircraft on a certain route depends on the mass m (in tonnes) of freight carried, according to the formula

$$C = 1000 + \frac{600m^2}{1+m}.$$

Copy and complete the following table, and hence draw on squared paper a graph of the function $f: m \to C$ for values of m from 0 to 5 inclusive.

m	0	1	2	3	4	5
$600m^2$	0	600	2400			
$1+m$	1	2	3			
$600m^2/(1+m)$	0	300				
C	1000	1300				

The airline charges £750 per tonne for freight. By drawing a further suitable straight line graph on the same diagram, find the mass of freight which must be carried on a flight to enable the airline to show a profit on that flight.

(MEI)

3 A missile is projected over the sea in such a way that the height h km above the sea and the distance d km that it has travelled horizontally are connected at any instant by the equation $25h = 40d - 4d^2$. Calculate

a) the height of the missile when it has travelled 6 km horizontally
b) the horizontal distance from the starting point when it is at a height of 3 km
c) the distance from the starting point when it strikes the water
d) the angle of elevation of the highest point reached, measured from the starting point.

4 The pilot of an aircraft wishes to fly from A to B, 240 kilometres away on a bearing of $270°$. The airspeed of the plane is 300 km/h and the wind is blowing from south to north at 30 km/h.

a) Calculate the course on which the pilot should fly, giving your answer to the nearest degree.
b) If the pilot sets his course due west, find his shortest distance from B.

5 A group of friends went into a restaurant for a meal. There were four courses on the menu and each of the party had just two of the four. Some had fish and a sweet, some had meat and a sweet, and some had meat followed by cheese and biscuits.
Let $\mathscr{E} = \{$The group of friends$\}$, $F = \{$Those who had fish$\}$, $M = \{$Those who had meat$\}$, $S = \{$Those who had a sweet$\}$, $C = \{$Those who had cheese and biscuits$\}$.

a) Explain in words i) $F \cap C = \phi$, ii) $M' = F$, iii) $C \subset M$, iv) $F \subset S$, v) $S' = C$.
b) If six of the party had meat, five had fish and four had cheese and biscuits, find how many there were in the party altogether.

6 The table shows how the power in megawatts (MW) supplied by a power station varies throughout a typical day.

Time	0000	0300	0600	0900	1200	1500	1800	2100	2400
Power (MW)	70	50	125	200	180	170	190	150	70

a) Illustrate these figures by drawing a graph. Estimate, to one significant figure, the rate in MW/hour at which the power supplied is decreasing at noon.
b) Use the trapezium rule with the given points to estimate the area under the graph, which gives the total energy supplied by the power station (in megawatt hours).
c) The average power for any period of time is found by dividing the total energy for that period by the number of hours. Estimate the average power supplied between 0600 and noon. (OXFORD)

Index

Note This is not a full index as the major topics listed on page 5 have been excluded.